VOLUME SIX

ADVANCES IN CHEMICAL POLLUTION, ENVIRONMENTAL MANAGEMENT AND PROTECTION

Wastewater Treatment and Reuse – Lessons Learned in Technological Developments and Management Issues

VOLUME SIX

ADVANCES IN CHEMICAL POLLUTION, ENVIRONMENTAL MANAGEMENT AND PROTECTION

Wastewater Treatment and Reuse – Lessons Learned in Technological Developments and Management Issues

Edited by

PAOLA VERLICCHI

University of Ferrara, Italy

ELSEVIER

ACADEMIC PRESS

An imprint of Elsevier

Academic Press is an imprint of Elsevier
50 Hampshire Street, 5th Floor, Cambridge, MA 02139, United States
525 B Street, Suite 1650, San Diego, CA 92101, United States
The Boulevard, Langford Lane, Kidlington, Oxford OX5 1GB, United Kingdom
125 London Wall, London, EC2Y 5AS, United Kingdom

First edition 2020

ISBN: 978-0-323-85684-3
ISSN: 2468-9289

For information on all Academic Press publications
visit our website at https://www.elsevier.com/books-and-journals

Publisher: Zoe Kruze
Acquisitions Editor: Jason Mitchell
Editorial Project Manager: Shellie Bryant
Production Project Manager: Denny Mansingh
Cover Designer: Victoria Pearson

Typeset by SPi Global, India

Contents

Contributors ix
Series editor's preface xi
Preface xiii

1. Treatment options for the direct reuse of reclaimed water in developing countries **1**
Tamires Lima Da Silva, Rodrigo Máximo Sánchez-Román, João Gabriel Thomaz Queluz, and Talita Aparecida Pletsch

1. Introduction 2
2. Availability of wastewater for the direct reuse of reclaimed water in developing countries 5
3. Wastewater irrigation in developing countries 9
4. Benefits and risks of the direct reuse of reclaimed water for irrigation in developing countries 12
5. Low-cost treatment options for the direct reuse of reclaimed water for irrigation in developing countries 14
6. Constructed wetlands for wastewater treatment 19
7. Combination of biofiltration and SODIS 23
8. Conclusions 27

Acknowledgment 28
References 28

2. Water reuse in India: Current perspective and future potential **33**
Arun Kumar and Kirti Goyal

1. Introduction 34
2. Potential for reclaimed water reuse in India 40
3. Applications of treated wastewater in India 45
4. Opportunities for wastewater reuse in India 49
5. Conclusions 60

References 61

3. Water reuse practices, solutions and trends at international **65**
Sylvain Donnaz

1. Water Reuse Practices Worldwide: "Not if, but how" 66
2. On the rise of water reuse worldwide 68

3. A brief review of water reuse technologies and solutions 72
4. The case of aquifer recharge in the value chain 77
5. Pioneering References and Flagship Water Reuse Projects around the World 84
6. Horizon 2040 of water reuse 95
Acknowledgments 100
References 100

4. Impact of the use of treated wastewater for agricultural need: Behavior of organic micropollutants in soil, transfer to crops, and related risks 103

Lapo Renai, Francesca Tozzi, Leonardo Checchini, and Massimo Del Bubba

1. Introduction 104
2. Use of WWs in agricultural production: Aim and consequences 106
3. Behavior of OMPs in soil 118
4. Transfer of organic micropollutants in edible crops 121
5. Concluding remarks 128
References 129

5. Environmental risks of sewage sludge reuse in agriculture 137

Elisabet Marti, Victoria Osorio, Marta Llorca, Lidia Paredes, and Meritxell Gros

1. Introduction 138
2. Overview of CECs in sewage sludge 139
3. Environmental risks associated with sewage sludge reuse 143
4. Conclusions 168
Acknowledgments 169
References 169

6. Modeling tools for risk management in reclaimed wastewater reuse systems: Focus on contaminants of emerging concern (CECs) 181

Riccardo Delli Compagni, Marco Gabrielli, Fabio Polesel, Andrea Turolla, Stefan Trapp, Luca Vezzaro, and Manuela Antonelli

1. Introduction 182
2. Reclaimed water reuse system: Boundaries and characteristics 185
3. Modeling reclaimed wastewater reuse systems for agricultural irrigation 190
4. Case studies 207
5. Conclusions 213
References 213

7. Feasibility evaluation in reclaimed water reuse projects through the analysis of some case studies 221

Paola Verlicchi and Giacomo Zanni

1. Introduction 222
2. Feasibility evaluation in agricultural and recreational reuse of reclaimed water 224
3. Decision-support models supporting reuse project evaluation 239
4. Final considerations 241

Appendix: Basic concepts in the evaluation of the economic and technical feasibility in case of reclaimed water reuse projects. 241

References 249

Contributors

Manuela Antonelli
Department of Civil and Environmental Engineering (DICA), Politecnico di Milano, Milano, Italy

Leonardo Checchini
Department of Chemistry, University of Florence, Florence, Italy

Massimo Del Bubba
Department of Chemistry, University of Florence, Florence, Italy

Riccardo Delli Compagni
Department of Civil and Environmental Engineering (DICA), Politecnico di Milano, Milano, Italy

Sylvain Donnaz
Suez International, Rueil-Malmaison, France

Marco Gabrielli
Department of Civil and Environmental Engineering (DICA), Politecnico di Milano, Milano, Italy

Kirti Goyal
Department of Hydro and Renewable Energy, Indian Institute of Technology, Roorkee, India

Meritxell Gros
Catalan Institute for Water Research (ICRA); University of Girona (UdG), Girona, Spain

Arun Kumar
Department of Hydro and Renewable Energy, Indian Institute of Technology, Roorkee, India

Tamires Lima Da Silva
Department of Rural Engineering, School of Agronomic Science—São Paulo State University, Botucatu, São Paulo, Brazil

Marta Llorca
Institute of Environmental Assessment and Water Research (IDAEA-CSIC), Barcelona, Spain

Elisabet Marti
Food Microbial Systems, Agroscope, Bern, Switzerland

Victoria Osorio
Beta Technological Centre (TECNIO Network), University of Vic-Central University of Catalonia, Barcelona, Spain

Lidia Paredes
Beta Technological Centre (TECNIO Network), University of Vic-Central University of Catalonia, Barcelona, Spain

Talita Aparecida Pletsch
Federal Institute of Education, Science and Technology of Espírito Santo, Campus Montanha, Montanha, Espírito Santo, Brazil

Fabio Polesel
DHI A/S, Hørsholm; DTU Environment, Technical University of Denmark, Kongens Lyngby, Denmark

João Gabriel Thomaz Queluz
Institute of Geosciences and Exact Sciences—São Paulo State University, Rio Claro, São Paulo, Brazil

Lapo Renai
Department of Chemistry, University of Florence, Florence, Italy

Rodrigo Máximo Sánchez-Román
Department of Rural Engineering, School of Agronomic Science—São Paulo State University, Botucatu, São Paulo, Brazil

Tamires Lima Da Silva
Department of Rural Engineering, School of Agronomic Science—São Paulo State University, Botucatu, São Paulo, Brazil

Francesca Tozzi
Council for Agricultural Research and Economics, Research Centre for Vegetable and Ornamental Crops, Pistoia, Italy

Stefan Trapp
DTU Environment, Technical University of Denmark, Kongens Lyngby, Denmark

Andrea Turolla
Department of Civil and Environmental Engineering (DICA), Politecnico di Milano, Milano, Italy

Paola Verlicchi
Department of Engineering, University of Ferrara, Ferrara, Italy

Luca Vezzaro
DTU Environment, Technical University of Denmark, Kongens Lyngby, Denmark

Giacomo Zanni
Department of Engineering, University of Ferrara, Ferrara, Italy

Series editor's preface

Let me introduce Volume 6 of the series entitled "Wastewater Treatment and Reuse—Lessons Learned in Technological Developments and Management Issues," edited by Paola Verlicchi. This is a follow-up and complementary reference to Volume 5 on "Wastewater Treatment and Reuse—Present and Future Perspectives in Technological Developments and Management Issues" edited by Paola as well on the same book series.

As a reminder, let me recall a few sentences from the preface to Volume 5. The book did describe different aspects of water reuse with comprehensive information, such as the European perspective; its use in agriculture, nature-based, and hybrid decentralized solutions; several well-known case studies of California, Singapore, and Australia; and the role of water reuse in a circular economy. Volume 6 is different but complementary to Volume 5, being more technologically oriented including management recommendations. It comprises seven chapters. The first one on treatment options for the direct reuse of reclaimed water in developing countries reports low-cost technologies such as the application of solar energy for disinfection as well as the very popular and eco-friendly approach of constructed wetlands. Other chapters describe the state-of-the-art water reuse in India as well as water reuse practices. The impact of the use of treated wastewater for agriculture with emphasis on organic contaminants in soil, their plant uptake, and related risks including sewage sludge reuse in agriculture is also presented. The last two chapters do relate modeling tools for risk management in reclaimed wastewater reuse with the focus on contaminants of emerging concern (CECs) and on the feasibility of reclaimed water reuse projects. Good case studies from Mexico, Jordan, Tunisia, Pakistan, and Ethiopia, among others, are being reported that show in practice how reclaimed wastewater reuse can be effectively implemented.

In the preface to Volume 5, I did mention that water reclamation and reuse is a key factor for a sustainable water resource management. Wastewater reuse can be of help to different sectors: irrigation for agriculture, industrial activities, potable demands, and city uses, among others. I can also add that although water treatment and reuse in Europe, the United States, Australia, and Singapore is under the radar of public and private sectors, it is not the case in other parts of the world, i.e., developing countries. I do believe that both Volumes 5 and 6 of this series would be of great help to increase and improve water treatment and reuse practices at a global level.

I thank all authors for preparing their timely, practical, and easily readable chapters. Volumes 5 and 6 are highly recommended for water managers; hydrologists; agricultural, chemical, and environmental engineers; economists working in the field of water quality; and newcomers and graduate students who wish to learn more on water treatment and reuse. To this end, I would like to add that both volumes include very clear and simple tables, depicting a lot of information and making these volumes useful and practical handbooks on water treatment and reuse. Both the volumes include safety measures and decision support tools that help solve the problems associated with water treatment and reuse.

Enjoy this wonderful and valuable reading of both comprehensive books on water treatment and reuse. Unfortunately when completing this Preface during this September 2020 we are still under the Covid-19 outbreak. Take care, use face masks, and stay safe and healthy!

Damià Barceló
ICRA-Catalan Institute for Water Research, Girona, Spain
and IDAEA-CSIC, Barcelona, Spain

Preface

The reuse of reclaimed water (RRW) has been the subject of many past studies and is still a topic of great concern today. The practice of RRW is in line with the European Green Deal, which represents a roadmap of actions to "boost the efficient use of resources by moving to a clean, circular economy and restore biodiversity and cut pollution" (https://ec.europa.eu/info/strategy/priorities-2019-2024/european-green-deal_en).

In the past RRW studies, population growth, climate change, severe and prolonged drought periods, water scarcity, and freshwater preservation and conservation were the main driving forces which promoted the worldwide development of RRW projects. Most efforts focused on the selection of the most adequate treatments and on the evaluation of their investment and operational costs. National quality standards, or alternatively standards defined by other countries (among them Israel and California), as well as the guidelines defined by the U.S. Environmental Protection Agency[1] and the World Health Organization[2] were adopted in selecting adequate technologies and sizing treatment steps.

More recently, new issues have attracted the attention and concern of scientists and practitioners: the occurrence in (waste)water of compounds of emerging concern (CECs), emerging pathogens as well as antibiotic-resistant bacteria and genes (ARB and ARG); their (often modest and/or scarce) removal achieved by conventional technologies; and their potential negative effects once in the environment, including the risk of entering the food chain. Despite the fact that CECs, ARB, and ARG are not regulated, efforts have been made and are still being made by scientists to identify the best available technologies and to optimize the processes to guarantee the effective removal of the key parameters.[3]

As a result, in recent years, RWW projects have started to include in their analysis the occurrence of a selection of emerging contaminants and the evaluation of their (expected) removal by the adopted treatment trains, the assessment of the risk to soil and crops posed by their residues in the treated effluent if directly reused for irrigation.

In this context, this volume aims to provide a snapshot of recent experiences of RRW in developing and developed countries; Managed Aquifer Recharge (MAR) coupled with water reuse; risk assessment in the case of the reuse of reclaimed water and treated sludge; and technical and economic feasibility studies.

In particular, the reuse of wastewater in developing countries is the subject of Chapter 1, which focuses on the main difficulties as well as the efforts required to guarantee safe water quality for rural needs. Experience in the direct reuse of water in India is outlined in Chapter 2 in terms of current perspectives and future potential. Chapter 3 deals with indirect potable reuse and develops the issue of MAR coupled with water reuse. The chapter also discusses the most adequate water treatment trains in this sector and presents some case studies of MAR around the world.

Risk assessment due to micropollutant residues in treated effluents and also in treated sludge spread onto rural land as amendments are the new emerging issues dealt with in Chapters 4–6. In particular, Chapter 4 carries out an analysis of the behavior of some organic micropollutants once in the soil due to irrigation with reclaimed water on the basis of their properties, the soil characteristics, and their transfer to edible crops. Chapter 5 deals with the environmental risks posed by the practice of sewage sludge reuse in agriculture due to the occurrence of different groups of CECs: pharmaceuticals and personal care products, microplastics, and engineered nanomaterials, as well as ARB and ARG. It also discusses the resulting risks of spreading sludge in rural areas to the soil, aquifers, and surface water, as well as to crops and human health. Finally, it suggests measures to reduce such risks. Chapter 6 suggests how to model the complex reclaimed water reuse system in order to fully understand the main processes that take place during wastewater treatment in the distribution network and once the treated water reaches the soil and crop.

And the last chapter focuses on the importance of carrying out an economic feasibility analysis of a reuse project. It discusses case studies which have evaluated cost-benefit analysis at different levels and it focuses on the different methods adopted for the estimation of the benefits identified for the specific cases (including the contingent valuation method and the travel cost method).

My wish is for the reader to be able to find some hints in this volume to help him/her develop water reuse projects on the basis of the lessons learned worldwide that are presented and discussed here and thus pursuing the objective of the sixth goal of the 2030 Agenda for Sustainable Development Goals (SDGs) which explicitly asks for a reduction in water pollution and for an improvement in ambient water quality, addressing human health, environmental protection, and economic activities.

To conclude, I would like to acknowledge all the colleagues who immediately agreed to join this editorial project and who have shared their know-how and skills with the scientific community, promoting the culture of the reuse of reclaimed water. A special thanks to Shellie Bryant and Jason Mitchell for their valuable and friendly collaboration right from the first contact at the beginning of this editorial adventure.

PAOLA VERLICCHI
Ferrara, August 2020

References

1. U.S.EPA. *Guidelines for Water Reuse—EPA/600/R-12/618*; September 2012.
2. WHO, UNEP. Excreta and Greywater in Agriculture. WHO Guidelines for the Safe Use of Wastewater, Excreta and Greywater [Internet]. 2006;vol IV:204. Available from: http://whqlibdoc.who.int/publications/2006/9241546832_eng.pdf.
3. Rizzo L, Gernjak W, Krzeminski P, Malato S, McArdell CS, Perez JAS, et al. Best available technologies and treatment trains to address current challenges in urban wastewater reuse for irrigation of crops in EU countries. Sci Total Environ [Internet]. 2020;710:136312. Available from: https://doi.org/10.1016/j.scitotenv.2019.136312.

CHAPTER ONE

Treatment options for the direct reuse of reclaimed water in developing countries

Tamires Lima Da Silva[a,*], Rodrigo Máximo Sánchez-Román[a], João Gabriel Thomaz Queluz[b], and Talita Aparecida Pletsch[c]

[a]Department of Rural Engineering, School of Agronomic Science—São Paulo State University, Botucatu, São Paulo, Brazil
[b]Institute of Geosciences and Exact Sciences—São Paulo State University, Rio Claro, São Paulo, Brazil
[c]Federal Institute of Education, Science and Technology of Espírito Santo, Campus Montanha, Montanha, Espírito Santo, Brazil
*Corresponding author: e-mail address: tamireslsilva@gmail.com

Contents

1. Introduction 2
2. Availability of wastewater for the direct reuse of reclaimed water in developing countries 5
3. Wastewater irrigation in developing countries 9
   3.1 Mexico 9
   3.2 Jordan 10
   3.3 Egypt 10
   3.4 Tunisia 11
   3.5 Pakistan 11
   3.6 Ethiopia 12
4. Benefits and risks of the direct reuse of reclaimed water for irrigation in developing countries 12
5. Low-cost treatment options for the direct reuse of reclaimed water for irrigation in developing countries 14
   5.1 Application of solar energy for disinfection purposes 14
6. Constructed wetlands for wastewater treatment 19
7. Combination of biofiltration and SODIS 23
8. Conclusions 27
Acknowledgment 28
References 28

Abstract

Reclaimed water may be defined as wastewater treated for a specific purpose. To be safely reused as reclaimed water, wastewater must receive adequate treatment to meet the quality requirements designed to minimize negative environmental impacts and to ensure the protection of public health. For developing countries, treatment methods

Advances in Chemical Pollution, Environmental Management and Protection, Volume 6
ISSN 2468-9289
https://doi.org/10.1016/bs.apmp.2020.07.006

that are expensive and/or complex to operate are not ideal due to local technical, financial, and managerial limitations. Thus, low-cost wastewater treatment technologies are often the best fits for developing nations. Nonetheless, low-cost wastewater treatment solutions are usually unable to treat wastewater for potable reuse due to the high quality required for this purpose. However, these technologies can efficiently treat wastewater for other purposes, such as irrigation. Irrigation with wastewater that has often not been properly treated or that remains untreated is already a common practice in many developing countries. Besides being a non-conventional water and nutrient resource for agricultural production, irrigation with untreated wastewater represents an epidemiologic risk to farms, consumers, and surrounding communities. Considering the importance of the development of applicable wastewater treatment options for developing countries in order to provide safe direct reuse of reclaimed water, this chapter will present three low-cost treatment solutions based on solar energy disinfection, phytoremediation, and a combination of biofiltration and solar energy disinfection. The wastewater treatment systems were designed to treat wastewater to the microbiological quality threshold recommended by the World Health Organization (WHO) for unrestricted irrigation (fecal coliform $\leq$1000 MPN/100 mL).

Keywords: Wastewater reuse, Solar disinfection, Constructed wetland, Biofilter

1. Introduction

Once freshwater has been used for a beneficial or economic purpose, it is usually released as wastewater into natural watercourses, either treated or untreated.[1] In some developing countries, it is estimated that >95% of all wastewater produced is discharged untreated into the environment.[2] Downstream of untreated wastewater discharge, effluent may be extracted and reused, which characterizes the process of indirect reuse.[3] Nevertheless, if the wastewater is treated and used for a specific purpose, such as for agricultural and landscape irrigation, aquaculture, industrial processes, groundwater recharge, urban use, environmental regeneration, or potable water supply, such reuse is considered direct and the treated wastewater can be defined as reclaimed water.[3]

The direct reuse of reclaimed water in developing countries presents many challenges. In the *2017 United Nations World Water Development Report*,[2] it is noted that issues related to lacking institutional and technical capacities, infrastructure, and financing for treatment contribute to the discharging of untreated wastewater into the environment in many developing nations. Consequently, the United Nations (UN) established target 6.3 in its *2030 Agenda for Sustainable Development*, which reads as follows:

"By 2030, improve water quality by reducing pollution, eliminating dumping and minimizing release of hazardous chemicals and materials, halving the proportion of untreated wastewater and substantially increasing recycling and safe reuse globally."[2] This target highlights the eminent need for technological options to adequately treat wastewater, thereby converting untreated wastewater into reclaimed water.

Though developing countries share some traits with respect to wastewater discharge and reclamation, there are institutional, technical, and economic differences among them. In the technical report, *Water Reuse in the Arab World: From Principle to Practice—Voices from the Field*,[4] it is noted that approximately 55% of the wastewater produced in the Middle East is treated, though only ~15% is reused for agriculture, landscape irrigation, groundwater recharge, environmental protection, and industrial cooling. This report shows that, overall, the development and implementation of water reclamation strategies across Middle Eastern countries is limited by the expenses of wastewater treatment and conveyance infrastructure, low pricing of irrigation water, insufficient economic analysis of wastewater treatment infrastructure projects, difficulty in generating financial incentives, and the distrust of farmers, who usually prefer other water resources, due the social stigma associated with the use of reclaimed water.[4]

In Latin America and the Caribbean, it is estimated that approximately 30% of wastewater is treated to some degree. Direct reuse of reclaimed water in this region is negatively affected by poorly developed legislation, a lack of regulatory mechanisms and policies to allow for gradual enhancement, technological selection criteria based on high-cost treatment solutions, dependence on conventional financing, and regulations that limit the reuse of reclaimed water. For instance, in the Province of Cordoba in Argentina, the legislation regarding the treatment of wastewater implemented in 2015 required that the discharge of wastewater treatment plants (WWTPs) into bodies of water should never exceed a concentration of $10\,mg\,N\,L^{-1}$ in the effluent, meaning that the average designed effluent value would have to be $\leq 3\,mg\,N\,L^{-1}$—a standard that few WWTPs in the world can meet.[5]

According to the *Policy Guidance Manual on Wastewater Management with a Special Emphasis on Decentralized Wastewater Treatment Systems (DEWATS)*[6] released by the UN Economic and Social Commission for Asia and the Pacific, the Asian and Pacific regions have limited sanitation coverage. Thus, innovative financing mechanisms supported by the Asia Development Bank, such as operational support funding in Pakistan, revolving credit pool schemes in Sri Lanka, viability gap funding in Kyrgyzstan, and output-based aid in

Nepal, have been provided to incentivize investments into sanitation and wastewater management systems.[7] Nevertheless, a lack of awareness of the impacts of untreated wastewater, low willingness to pay for treatment, and misconceptions of the potential benefits of such treatment are challenges that still need to be overcome by planners and policymakers.

African countries face enormous challenges in providing appropriate wastewater treatment and reuse. The management of wastewater is limited by overloaded or nonfunctioning WWTPs,[8] lacking power supplies, outdated regulations, and differing national policies.[9] Investments into the maintenance and construction of new wastewater treatment facilities have been supported by the African Development Bank, which provided funding to the Nakivubo WWTP in Uganda[10]; public–private sector partnerships are also important means of funding that have been used for African countries. The WWTP in Durban, South Africa and the WWTP in New Cairo, Egypt are examples of successful public–private partnerships.[11]

Reclaimed water must receive adequate treatment in order to meet the quality requirements designed to minimize negative environmental impacts and ensure the protections of public health. For developing countries, treatment methods that are expensive and/or complex to operate are not ideal due to national technical, financial, and managerial limitations. The low-cost treatment solutions that are currently available and enable the extraction of energy and nutrients from wastewater may be a good fit for developing countries.[12–14] Nevertheless, low-cost wastewater treatment technologies are usually not able to treat wastewater for potable reuse because of the high quality required for this purpose. However, these technologies can efficiently treat wastewater for other purposes, such as irrigation reuse.

For many farmers in developing countries, it is a common practice to indirectly reuse wastewater, raw or diluted, to irrigate crops, especially vegetables, in urban and peri-urban areas.[15,16] Despite the advantages of the high nutrient contents of wastewaters, such water resources can often be contaminated with pathogens and chemical agents that pose significant risks to the health of the farmers, their surrounding communities, and to consumers.[15,17,18] Sewage treatment in developing countries is also limited and the centralized or decentralized collection and treatment systems cannot always generate wastewater that meets the quality recommended for irrigation purposes, due to their poor operation and/or maintenance, as well as overflow during storm events.[2,19] Therefore, treated wastewater will sometimes requires additional treatment prior to reuse in irrigation. Considering the importance of developing applicable wastewater treatment options for developing countries, in order to allow for the safe direct reuse of reclaimed water, three low-cost treatment solutions based on solar energy disinfection, phytoremediation, and the combination of biofiltration and solar energy disinfection are presented in this chapter. The wastewater treatment systems were designed to treat wastewater to the microbiological quality threshold recommended by the World Health Organization (WHO) for unrestricted irrigation (fecal coliform $\leq$1000 MPN/100 mL).

2. Availability of wastewater for the direct reuse of reclaimed water in developing countries

The UN designations, "developed" and "developing country," are used for areas in the UN system for statistical convenience, and they are not intended to express a judgment about the stage reached by a particular country or area in the developmental process, although this distinction remains relevant to the Sustainable Development Goals. According to UN standard country or area codes, there are 181 countries designated as "developing countries."[20] Information regarding the quantity of wastewater produced or treated at the national scale in developing countries may often be limited or outdated. By examining the availability of wastewater data for the 181 developing countries in the AQUASTAT Main Database of the UN Food and Agriculture Organization (FAO),[21] it is clear that these data are limited and/or outdated regarding the quantity of wastewater produced or treated. Data are available for only 79 countries (44%) on the wastewater produced, and 63 countries (35%) on the quantity of treated wastewater (Fig. 1).

Real data on the generation and treatment of wastewater at the national scale is essential for researchers, policy makers, and institutions to develop national action plans aimed at the direct reuse of reclaimed water. However, the data available in the AQUASTAT Main Database for some developing countries reveals that information on the generation and treatment of wastewater is not recent; in fact, ~47% of the information on produced wastewater and 46% on treated wastewater at that national scale were collected before 2010 (Fig. 2). In such cases, the values may not represent the current situation, since the volume of wastewater may be increasing due to population growth, urbanization, economic development, and improvements in living conditions (especially plumbing and municipal waterworks).[8,9,22]

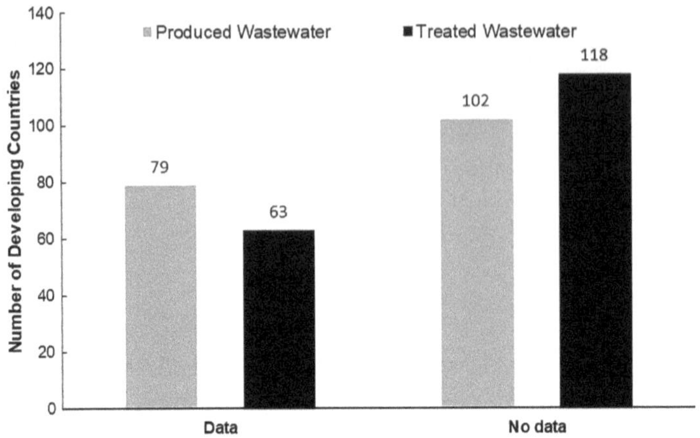

Fig. 1 Availability of wastewater production and treatment data for developing countries.

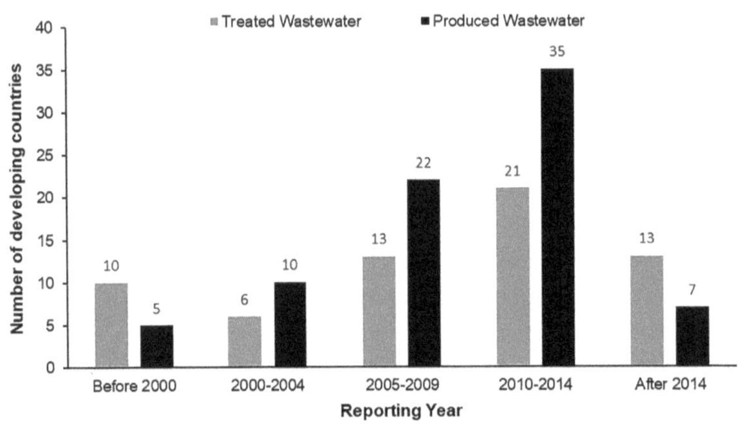

Fig. 2 Timeline for data available on wastewater production and treatment in developing countries.

Despite the importance of wastewater treatment in preventing water pollution, in many developing countries, the wastewater produced is still released into the environment without treatment. The data analyzed from AQUASTAT reveal that only 30% of the developing countries evaluated have information available for both the generation and treatment of wastewater. For these countries, based on the produced and treated amounts, an average of just 37% of the wastewater produced is actually treated (Fig. 3).

Irrigation is a key factor for ensuring food security in developing countries, and it is worrisome that, for some countries, the availability of water for

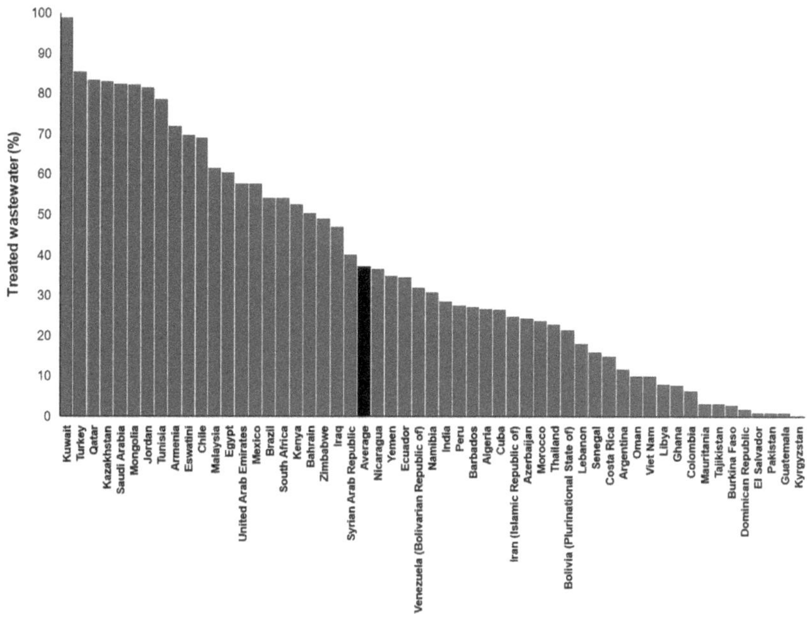

Fig. 3 Percentage of treated wastewater in developing countries.

agricultural irrigation is extremely limited.[23,24] In this context, the volume of available wastewater at the national scale, if adequately treated, can be transformed into a valuable source of water and nutrients for crop production, thereby improving food security. Fig. 4 shows the amount of wastewater produced in developing countries; a large volume of wastewater, $>10.30 \times 10^9$ m^3/year, is produced in countries across Asia. The least amount of data is available for the continent of Africa, which is a serious issue that impacts the management and planning of wastewater treatment for direct reuse in agricultural irrigation. In sharp contrast, in Latin America and the Caribbean, very few countries present lack of data.

In spite of the fact that data on the generation and treatment of wastewater in developing countries are lacking or outdated, when the volume of treated wastewater (Fig. 5) is compared to the amount of wastewater produced (Fig. 4), it is clear that in these countries, the majority of wastewater is released untreated into the environment. For instance, the observed volume of wastewater produced in India ranges from 10.30 to 15.45×10^9 m^3/year (Fig. 4), while the volume of treated wastewater ranges from 2.6 to 4.42×10^9 m^3/year (Fig. 5). Considering these ranges, it has been estimated that less of 30% of all of the wastewater produced in India is reported to be

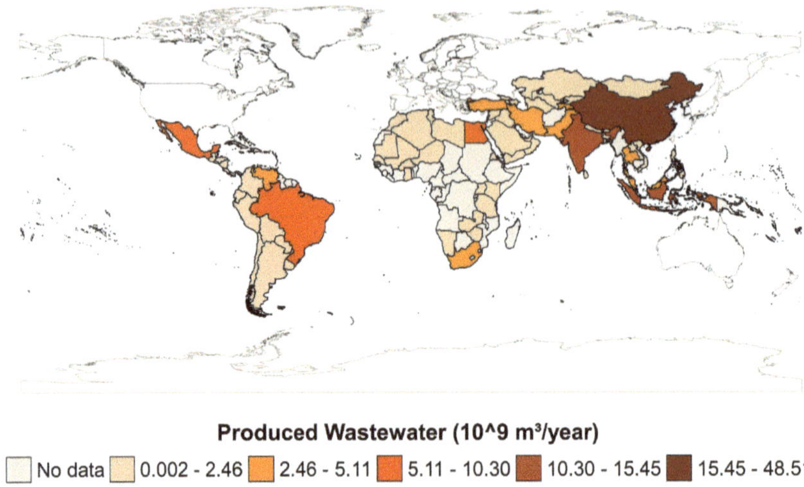

Fig. 4 Volume of wastewater produced in developing countries.

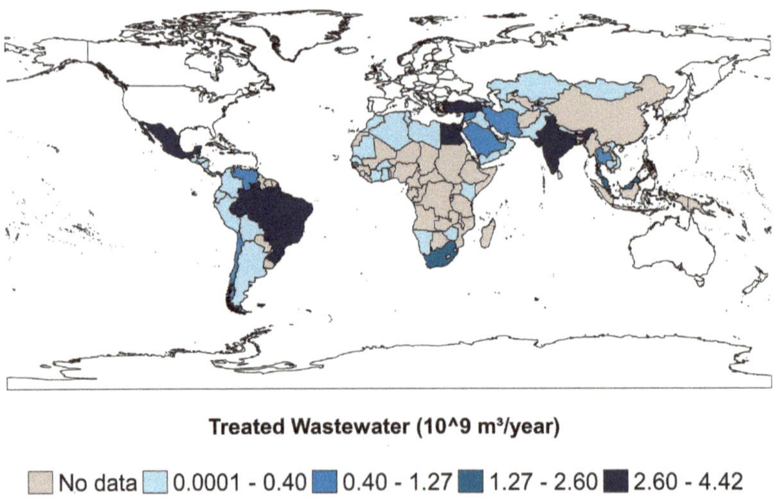

Fig. 5 Volume of treated wastewater in developing countries.

treated. It is important to highlight that the discharge of untreated wastewater can be harmful to human health, generating outbreaks of food-, water-, and vector-borne diseases, and may also have negative environmental effects, including water and soil pollution, loss of ecosystem services, and adverse repercussions on economic activities, such as the increased cost of water treatment downstream.[2]

Water resources and wastewater management can both be affected by urbanization, industrialization, population growth, climate change, and changes in consumption patterns. Population growth, urbanization, industrialization, and increased consumption increase the demand for water and consequently lead to increases in the generation of wastewater,[2,8,9,22] demonstrating that investing into wastewater treatment for the direct reuse of reclaimed water in developing countries is a great opportunity to capitalize upon reclaimed water as a valuable alternative water resource. For this reason, accurate information on wastewater production in developing countries is indispensable for improving wastewater management for water reclamation.

3. Wastewater irrigation in developing countries

The reuse of wastewater for irrigation is a common practice in developing countries. A comprehensive research report released by the International Water Management Institute (IWMI) in 2008 described the drivers and characteristics of the agricultural reuse of wastewater in developing countries.[22] Through a cross-country analysis of 53 cities, the results showed that the main drivers of wastewater reuse for irrigation in developing countries are: (a) a lack of cheaper water sources, (2) market and urban food demand incentives advantaging food production close to cities where water sources are frequently polluted, thus increasing urban water demand and its return into the environment, and (3) the contamination of traditional water sources. Several examples of the used of wastewater for irrigation in developing countries are presented below.

3.1 Mexico

One unique example of wastewater irrigation occurs in the Mezquital Valley of Mexico, where an extensive network of irrigation canals conducts 4.49 Mm^3 day^{-1} of wastewater to irrigate maize, lucerne, fodder oat, rape, zucchini, cauliflower, chili pepper, and ryegrass crops. Irrigation with wastewater was officially permitted in 1912, and the irrigated area has been extended as wastewater discharge has increased, reaching approximately 90,000 ha at present.[25] The nutrient and organic contents of the wastewater have benefited crop production in the Mezquital Valley; the crop yields achieved on farms irrigated with wastewater are usually higher than rain-fed and well-water-irrigated farms. Until 2015, only untreated wastewater was used for irrigation, but in 2016, a large WWTP began operating in the region.[26] However, the long-term irrigation with untreated wastewater has resulted in public health risks. An epidemiologic study conducted in the Mezquital Valley showed that communities exposed to wastewater had a higher one-week prevalence of diarrhea (10%) compared to unexposed communities (5%).[18]

3.2 Jordan

Jordan has renewable water resources of only 145 cubic meter per capita per year, so the demand for water often exceeds the available supply.[27] Thus, wastewater has been used as an unconventional water resource to reduce the gap between the supply and demand. In 1998, the Wastewater Management Policy was established by the Ministry of Water and Irrigation; this policy emphasized that treated wastewater must be considered as a water resource and part of the water budget, such that wastewater should not be disposed and crops irrigated with wastewater or blended water should be monitored.[28] In 1970, Jordan's first WWTP was established, and by 2012, there were 26 such facilities. A large portion of the treated wastewater is used for irrigation. The As-Samra WWTP, which is the largest in Jordan, has been used to irrigate vegetables, bananas, grapes, citrus, and certain stone fruits, in the Jordan Valley. The main concerns regarding the use of reclaimed water for irrigation in the Jordan Valley are high salinity and high chloride concentrations that can reduce the yields of sensitive crops.[29]

3.3 Egypt

Rainfall in Egypt is very irregular, annual rainfall ranges between a maximum of about 200 mm in the northern coastal region to a minimum of nearly zero in the south,[30] this fact renders Egyptian agriculture heavily dependent upon irrigation. The reuse of wastewater for irrigation is an old practice in Egypt. Since 1900, sewage water has been used to irrigate crops in the village of El-Gabal El-Asfar (~4500 ha).[23] A more recent example of agricultural reuse of wastewater in Egypt is the El Berka WWTP and its reuse scheme. The plant was constructed in 1990 with primary treatment only, while secondary treatment was established in 1998. The facility has a total area of 42 ha and receives wastewater from approximately one million households. The treated wastewater is reused to irrigate lemons, cacti, *Cupressus sempervirens*, *Khaya senegalensis*, *Jatropha*, and *Jojoba* crops.[31] Under Egyptian law, the reuse of treated wastewater is not allowed for irrigating food and fibers. Nevertheless, a lack of investment in wastewater collection and treatment infrastructure in the past has resulted in surface water pollution. The discharge of raw or inadequately treated wastewater into the Nile River has led to the degradation of water quality in the Nile Delta region. For example, three agricultural reuse pump stations had to be shut down because of the high level of pollution in drainage waters.[23]

3.4 Tunisia

With the objective of taking pressure off of groundwater resources susceptible to saltwater intrusion, Tunisia began experimenting with wastewater reuse for agricultural irrigation in 1965, when citrus groves cultivated in La Soukra were irrigated with treated wastewater from the Cherguia WWTP. The reuse of wastewater is regulated by the 1975 Water Code, as well as by subsequent standards and decrees. Decree No. 89-1047 (issued on July 28, 1989) specifically regulates the reuse of wastewater in agriculture. The irrigation of vegetables that may be consumed raw is prohibited by the Water Code and reclaimed water is mainly used to irrigate fruit trees, industrial crops, forages, and landscaping. Additionally, specifications regarding precautions that must be taken to protect workers, residential areas, and consumers at risk of contamination have been established.[32,33]

3.5 Pakistan

Wastewater is a valuable resource for farmers in Pakistan due its high nutrient concentrations (e.g., nitrogen, phosphorus, and potassium, NPK). Wastewater irrigation can reduce the application of chemical fertilizers or even make additional applications unnecessary. Despite the potential for negative environmental and public health impacts, most domestic and industrial wastewaters in Pakistan are released into water bodies or natural drainages, and there remains no national strategy related to the management of wastewater for irrigation purposes.[34] In February 2000, the IWMI began an interdisciplinary study in the city of Haroonabad, located in the southern Punjab Province, where untreated urban wastewater had been used for agricultural irrigation for the past 35 years, mostly for vegetable crops. The results of this study showed that farmers that used wastewater for irrigation saved significantly on chemical fertilizer and irrigation water. However, wastewater analyses also revealed high concentration of helminth eggs and fecal coliform, presenting potential health risk to farmers, consumers, and surrounding communities.[17,34]

3.6 Ethiopia

As in most developing countries, polluted stream water in Ethiopia has been used to irrigate a variety of crops. In the capital city, Addis Ababa, wastewater has been released since the 1940s into streams flowing through rivers, being collected and indirectly reused for crop production. Farms supply the city with fruits and almost all vegetables crops (e.g., lettuce, Swiss chard, cabbage, spring onion, potato, beat root, etc.).[8] Besides the benefit to city residents in obtaining fresh vegetables, crop production with irrigated wastewater provides income for many small-scale farmers and traders. Weldesilassie and colleagues [15] studied wastewater reuse in crop production in the peri-urban areas of Addis Ababa, and observed that, for the majority of farm households, vegetable production with wastewater irrigation was a major source of income; for example, an average farm household using wastewater for irrigation gained an annual net income of Birr 4′448, of which ~70% was derived from the wastewater farm. Nevertheless, working on wastewater-irrigated farms significantly increased the frequency of illness, which could, in-turn, reduce the household income.

4. Benefits and risks of the direct reuse of reclaimed water for irrigation in developing countries

In many developing countries, untreated wastewater (raw or diluted) has been widely used for irrigation; besides being a source of water and nutrients for agriculture, this practice can have negative impacts on public health, crops, soils, and the environment. One of the most recognized benefits of the direct reuse of reclaimed water in agriculture is increases in crop productivity. A field study carried out at the College of Agriculture Farm in Nagpur, India evaluated the use of domestic wastewater for crop irrigation. Wheat, gram, palak, methi, and berseem were irrigated with groundwater and domestic wastewater and received the recommended nitrogen-phosphorus-potassium (NPK) dose of fertilizers. Irrigation with domestic wastewater improved the physicochemical properties of the soil, crop yields, and nutrient status. Slightly more NPK was available for the crops irrigated with wastewater, as compared to crops irrigated with groundwater (2.4% N, 14.1% P, and 1.2% K). Additionally, the use of domestic wastewater favorably influenced the crop production of grain and straw yields of wheat (*Triticum aestivum*), palak (*Spinacia oleracea*), and berseem (*Trifolium alexandrinum* L.), which were higher in the treatments irrigated with wastewater.[35]

Reclaimed water is an alternative source of water for agriculture, especially for developing countries facing water shortages, growing urban populations, and climate change. Due to water stress, the reuse of reclaimed water for irrigation has been designated as a vital part of an integrated water resources management approach aimed at narrowing the gap between supply and demand.[9,23,24] The direct reuse of reclaimed water for irrigation contributes to the prevention of water pollution. Once discharged into water bodies, wastewater can generate pollution due its physical, chemical, and biological characteristics. High nutrient concentrations (i.e., N, P and K) can accelerate the eutrophication freshwater resources and increasing loads of phosphorus have been reported in most of the largest lakes in Africa and Latin America.[36] In turn, eutrophication can lead to toxic algal blooms, impacting ecosystems services and decreasing biodiversity. The 2017 UN World Water Development Report, "*Wastewater: The Untapped Resource*," indicated that ~33% of all river stretches in Asia, Africa, and Latin America are severely polluted by pathogens (e.g., fecal coliform >1000 MPN/100 mL), which present serious risks to public health.[2] Moreover, Wen et al.[37] projected that the number of people affected by the organic pollution of rivers (biochemical oxygen demand, BOD >5 mg/L) will increase from 1.1 billion in 2000 to 2.5 billion by 2050, with developing countries being disproportionately affected, if wastewater treatments continue to be insufficient at keeping BOD concentrations below 5 mg/L.

The reuse of reclaimed water in agriculture is not exempt from negative impacts on soil and crops. Angin et al.[38] analyzed changes in soil contents after long-term irrigation with wastewater on the Erzurum Plain of Turkey. They observed an increase in salt content in wastewater-irrigated soils compared to non-wastewater-irrigated soils, likely as consequence of high electrical conductivity in the wastewater. Irrigation with highly saline water may lead to salinity problems that could affected crop productivity.[39] Additionally, the results of their study showed an increase in the accumulation of heavy metals (i.e., Cu, Fe, Mn, and Zn) in soils irrigated with wastewater, when compared to non-wastewater-irrigated soils, and high levels of heavy metals can result in crop toxicity.[39]

Agricultural reuse of untreated wastewater is very dangerous due to the presence of emerging contaminants (e.g., pesticides, personal care products, pharmaceuticals, drugs, etc.) and pathogens (bacteria, viruses, protozoa, nematodes, helminths, etc.).[40,41] In Ghana, researchers evaluated the contribution of wastewater irrigation to the transmission of helminths in soils among vegetable farmers.[42] It was observed that farmers working with wastewater irrigation were three times more likely to be infected with *Ascaris* sp. and hookworm than unexposed individuals, indicating the direct linkage between wastewater irrigation and helminth infection in exposed farmers.

In Saudi Arabia, researchers found that emerging contaminants can be uptaken by crops. A study conducted in the Al Hayer area identified 64 emerging contaminants in wastewater and 13 in crops (six pharmaceuticals and seven pesticides). Such emerging contaminants are accumulated and can pass up the *food chain (bioaccumulation), which is why they are a major concern for public health.*[43]

Due to the public health risks associated with the reuse of reclaimed water in agriculture, the WHO regularly publishes guidelines for the safe reuse of wastewater. The most recent set of guidelines was published in 2006. With respect to the safe reuse of wastewater for irrigation purposes, the WHO recommends that a microbiological quality of fecal coliform $\leq$1000 MPN/100 mL and helminth $\leq$1 eggs/L for unrestricted irrigation (crops that are eaten raw) of root crops.[44] Furthermore, the FAO has developed guidelines for wastewater reuse for agriculture; in the FAO's irrigation and drainage paper No. 47, "*Wastewater Treatment and Use in Agriculture*," it is recommended that wastewater be classified, based on physical and chemical parameters, into one of three degrees of restriction on use: none, slight to moderate, and severe.[39] In developing countries with considerable reuse potential or those in which reuse is currently practiced, national policies and standards must be created and the harmonization of relevant policies already implemented that interface with reclaimed water reuse must be promoted, rather than simply adopting international guidelines, since each country has unique institutional, technical, and economic specificities.

5. Low-cost treatment options for the direct reuse of reclaimed water for irrigation in developing countries

5.1 Application of solar energy for disinfection purposes

The germicidal action of solar energy on bacteria and other organisms was observed in 1877 by Downes and Blunt[45]; since then, many studies have been conducted to better understand this process. Solar disinfection (SODIS) occurs when pathogens are rendered inactivate by exposure to sunlight. The spectrum of solar radiation reaching the Earth's surface consists of ultraviolet (UV), visible, and infrared radiation. Solar disinfection occurs mainly due to the effects of UV radiation on the activity of microorganisms. The SODIS process is complex, with multiple steps occurring simultaneously. The deoxyribonucleic acid (DNA) of microorganisms is damaged by the direct absorption of light, which can injure DNA-generating mutations and/or stop DNA replication. At the same time, the formation of reactive oxygen species may contribute in the deactivation process. A synergetic effect has also been observed between high water temperatures (>45 °C) and the deactivation of microorganisms in the process of SODIS. Other factors that also may also affect the efficiency of SODIS include the amount of total suspended solids (TSS), turbidity, and seasonal variations in sunlight.[46]

Some solutions for wastewater disinfection are difficult to apply in developing countries and rural communities, either because there are expensive or because they require qualified workers to operate. Chlorine treatment is among the inexpensive options for wastewater disinfection; nevertheless, once chlorine is applied in wastewater treatment, it may have residual effects, due to reactions with natural organic matter, consequently forming carcinogenic halogenated disinfection by-products, such as haloacetic acids and trihalomethanes.[47,48] The bactericidal effect of solar radiation has been demonstrated in several studies, indicating its potential to treat wastewater.[13,49–51] Solar disinfection systems have low installation and maintenance costs, and are easy to use when compared to others disinfection systems. Thus, SODIS is a wastewater treatment option with wide applicability in regions lacking financial resources.

As cost-effective treatment solutions are more likely to be applied in developing countries, a SODIS reactor with a concrete base[50,51] is the first wastewater treatment option presented here, with the aim of achieving the water quality necessary for unrestricted irrigation. The SODIS reactor has a concrete base and is shaped like an inverted truncated cone with the following measurements: larger radius, 1.00 m, smaller radius, 0.25 m; height, 0.30 m (Fig. 6). This form was adopted to ensure that the walls of the reactor did not produce shadows on the effluent surface for the longest time possible.

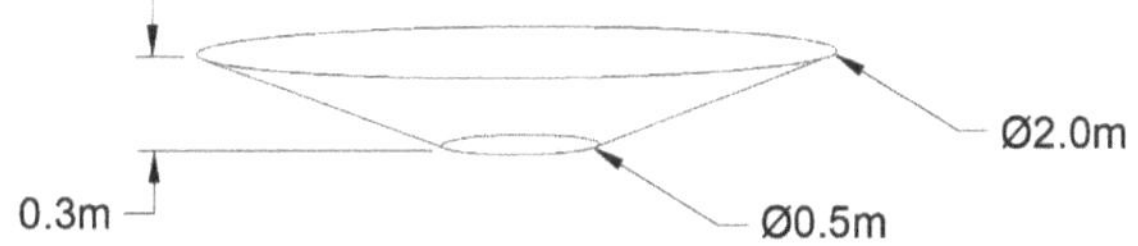

Fig. 6 Structural design of the solar disinfection (SODIS) reactor.

Fig. 7 Differently colored SODIS reactors.

Aiming to analyze the possible synergism between solar reflection and temperature in SODIS efficiency, three reactors with different colors: black, white, and unpainted (concrete) were tested (Fig. 7) by Queluz.[51] The study was carried out from March to April of 2013 at the School of Agronomic Sciences (FCA), Unesp, Botucatu-SP, Brazil (22°51′12″S, 48°25′45″W).[23] The SODIS efficiency was evaluated by analyzing *Escherichia coli* (*E. coli*), chemical oxygen demand (COD), wastewater temperature, global radiation, UV radiation, pH, and total suspended TSS. The bacterium, *E. coli*, was adopted as the fecal indicator bacteria (FIB) to detect the occurrence of fecal contamination.

Global solar radiation and UV radiation data were obtained from the weather station of the Department of Environmental Resources and Soil Sciences located on the FCA campus. The wastewater temperature was measured using a thermometer (Incoterm model: 7666.02.0.00). The parameters of *E. coli* (MPN/100mL), COD (mg/L), pH, and TSS (mg/L) were then analyzed according to the *Standard Method for the Examination of Water and Wastewater*[52] (Table 1).

Wastewater from the Botucatu WWTP in Botucatu, São Paulo, Brazil was directed to a septic tank with the aim of reducing the concentrations of solids. The hydraulic retention time of the septic tank was 24h after that the wastewater was conducted to the SODIS reactors (Fig. 8). In the reactors, a wastewater depth of 0.10 m was exposed to sunlight for three consecutive days. The wastewater samples were collected and preserved according to the Standard Methods 1060B and 1060C, respectively. During the 3 days, the

Table 1 Parameters analyzed and the Standard Methods adopted for examining wastewater.

Parameter	Standard method for the examination of water and wastewater
E. coli (MPN/100 mL)	9223B Enzyme substrate test with multi-well procedure
Total suspended solids (mg/L)	2540D Total suspended solids dried at 103–105 °C
Chemical oxygen demand (mg/L)	5220D Closed reflux, colorimetric method
pH	4500B Electrometric method

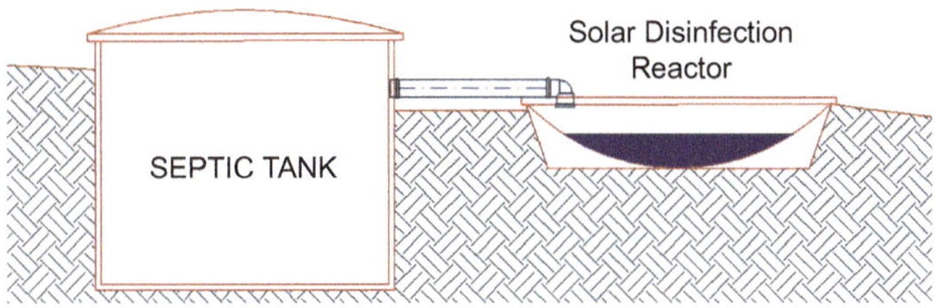

Fig. 8 Wastewater SODIS system.

first wastewater sample was collected at 8:00 a.m. (local time), then subsequent collections were performed every 5 h until 6:00 p.m. This procedure was performed six times for each reactor.

Based on the results of the accumulated global radiation dose ($MJ\,m^{-2}$), accumulated UV radiation dose ($MJ\,m^{-2}$), wastewater temperature (°C), COD (mg/L), TSS (mg/L), pH, and the application of mathematical models, the efficiencies of SODIS reactors in reducing *E. coli* concentrations were evaluated using regression analyses in the statistical software R (version 3.0.1). To compare the efficiencies of the treatments and the experimental replicates, the response variable used was the ratio between the initial population (*No*) and the remaining population (*N*) of *E. coli* after exposure to solar radiation ($y = N/No$). In this way, the values of the dependent variable will always be ≤ 1 and always >0. The function used to analyze the data and adjust the models was a logistical function (S-shaped function) with two parameters. Additionally, using linear models and adopting the logarithmic reduction of the *E. coli* population (reduced LOG = log (1/y)) as the response variable, 12 mathematical models were adjusted and compared.

The parameters adopted for comparison between models were the Akaike Information Criterion (AIC) and Bayesian Information Criterion (BIC). These parameters can be used to compare models in the same way as analyses of variance, wherein the model with the lowest values for these two parameters is what best explains the studied dataset. Taking into consideration that the UV fraction was equal to 5% and the methodology, the monthly mean incident UV radiation in Botucatu City was estimated.

The effluent from the septic tank had an arithmetic mean of 1.29×10^6 MPN *E. coli*/100 mL, COD of 655 mg/L, TSS of 180 mg/L, and pH = 6.17. The analyses of wastewater samples did not show significant differences between the reactors, and even the wastewater temperatures were similar; for example, the arithmetic mean temperature difference between the black and white reactor was only 2.06 °C. Considering the results for all three reactors, the mean values were: COD, 487 mg/L; TSS, 134 mg/L; pH, 7.11. Compared to the effluent from the septic tank, the reactors decreased the organic matter (COD) and concentration of solids (TSS) by ~25.6% and ~25.4%, respectively; in contrast, pH increased by 15.2%.

Using the remaining population of *E. coli* in relation to the global solar radiation dose and regression analysis, a mathematical model was developed for each reactor studied. However, the data obtained from the descriptive statistics indicated that the treatments did not show statistically significant difference at the 5% alpha-level (i.e., $p > 0.05$), indicating that the use of different colorations in solar reactors did not strongly affect the efficiency of SODIS. For this reason, the data from the three SODIS reactors were grouped and adjusted into a single model, as presented in Eq. 1:

$$\frac{N}{No} = \frac{\exp(2.017 - 0.2321RG)}{(1 + \exp(2.017 - 0.2321RG))}, \tag{1}$$

where N is the population of *E. coli* after the exposure of wastewater to solar radiation (MPN/100 mL), No is the initial population of *E. coli* in the wastewater (MPN/100 mL), and RG is the dose of solar global radiation received by the wastewater ($\mathrm{MJ\,m^{-2}}$).

The evaluation of the AIC and BIC parameters of the 12 linear models showed that the lowest values were associated in the model designated "X," suggesting that this model best fit the data. The significant variables of this model were UV radiation and the TSS concentration. The obtained Pearson's correlation coefficient (R^2) was 0.8277. Model "X" is presented in Eq. 2:

$$y = \frac{N}{No} = \frac{1}{10^{(1.254741+0.703179RUV-0.006492TSS)}}, \quad (2)$$

where N is the remaining population of *E. coli* after the wastewater exposure to UV radiation (MPN/100 mL); No is the initial population of *E. coli* on wastewater (MPN/100 mL); RUV is the accumulated dose of UV radiation ($MJ\,m^{-2}$); and TSS is the concentration of total suspended solids (mg/L).

Based on Eq. 2, the estimated incident UV radiation in Botucatu and six fixed concentrations of TSS (50, 100, 125, 150, 175, and 200 mg/L), the time required for solar UV radiation to disinfect wastewater with an average *E. coli* concentration of 1.29×10^6 MPN/100 mL to the quality level recommended by the WHO for unrestricted irrigation reuse ($\leq$1000 MPN/100 mL), may range from 2.82 to 7.19 days. It should be noted that TSS concentrations are not constant in wastewater, as assumed in the simulated scenarios. For conditions similar to those in this work, the remaining population of *E. coli* in wastewater after being exposed to solar radiation can be calculated by Eq. 2.

6. Constructed wetlands for wastewater treatment

Phytoremediation is the application of plants for the removal, degradation, and detoxification of contaminants, including organic matter, inorganic matter, suspended solids, trace organics, and pathogens.[53] Phytoremediation in constructed wetlands has been widely used as wastewater treatment option. Constructed wetlands used for wastewater treatment are systems that have been engineered to mimic natural wetland processes within a more controlled environment.[12]

According to Kivaisi,[12] the potential application of wetland technologies in developing countries is tremendous, since most developing countries are located in warm tropical and subtropical regions, characterized by intense microbiological activity that guarantee the high removal of pollutants, thus contributing to the performance of constructed wetland systems. Tropical and subtropical areas are known to sustain rich species diversity that could be used in wetlands, though the availability of land might be an issue in dense urban areas. Therefore, constructed wetlands are potentially well-suited to smaller communities.

The biological processes occurring in constructed wetlands vary and may affect different types of pollutants. Moreover, they can be used as secondary or tertiary treatments.[12] Constructed wetlands are nature-based and considered suitable for applications in developing countries with tropical and subtropical climates. For this reason, the second low-cost treatment option for the direct reuse of reclaimed water for unrestricted irrigation in developing countries is the constructed wetland system investigated by Queluz.[54]

The constructed wetland treatment system was composed of four rectangular fiberglass water tanks (100 × 35 × 31 cm) connected in a series, with an estimated storage capacity of 100 L and useful volume of 160 L. The water tanks were filled with gravel (Ø = 2.49.5 mm), with mean porosity of 51%, to a height of 30 cm, and then it was saturated with wastewater to a height of 25 cm. The effluent level in the last water storage tank (25 cm) was obtained and controlled using a discharge- and level-control device composed of PVC piping and screw connections installed in the last water tank. Using pieces of wood, a slope of 1.5% between each water tanks was produced; the first water tank was placed 6 cm from the ground, the second, 4 cm, the third, 2 cm, and the last one, at ground level, so that water could flow by gravity. The beds were vegetated with cattails (*Typha latifolia*), with four plants in each bed. The system received 50 L/day of wastewater, resulting in a nominal hydraulic retention time of 3.2 days (Fig. 9).

The constructed wetland system was tested at the FCA. According to the Köppen climate classification,[55] Botucaty City is classified as Cfa, humid subtropical climate with rains in the summer and drought in the winter that presents an average temperature of the hottest month above 22 °C, and the mean annual precipitation and mean annual temperature are 1509 mm and 21 °C, respectively. The cattails were collected in natural wetlands and then half-height pruned (≈50 cm) to be replanted in the constructed wetland in August of 2015. During the period of fixation of the cattails, the system was saturated with water and after the establishment of the macrophytes, the application of wastewater was initiated on November 1, 2015. Raw or diluted wastewater (50 L) was applied daily in the constructed wetland system from November 1, 2015 to April 29, 2016. Every 2 weeks, wastewater from the Botucatu WWTP (secondary treatment) was transported by a tank truck to the experimental site, where it was stored in a 1000 L plastic water storage tank. Over 60 days, different rates of wastewater were applied in three distinct phases: Phase 1, 50% wastewater +50% water (November 1 to December 28, 2015); Phase 2, 75% wastewater +25% water (December 29, 2015 to March 01, 2016); Phase 3, 100% wastewater (March 2 to April 29, 2016). Four samplings were carried out in which water samples were collected according to the *Standard Methods for Examination of Water and Wastewater. Wastewater* (raw or mixed with water). Samples were analyzed

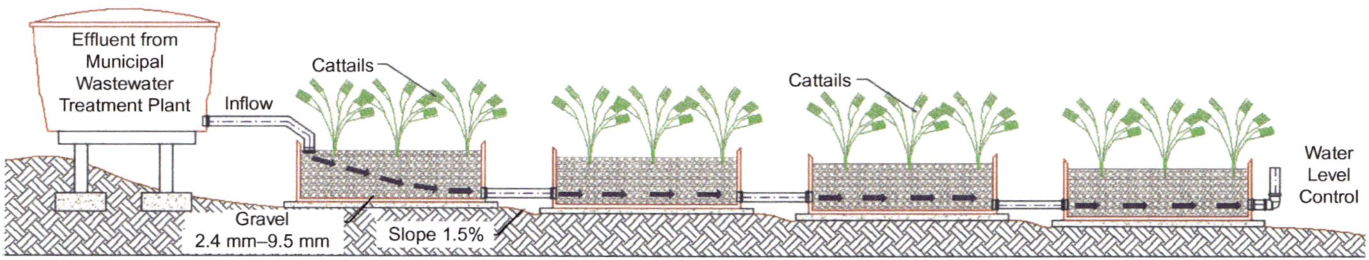

Fig. 9 Constructed wetland cultivated with cattails (*Typha latifolia*).

Table 2 Parameters analyzed and the Standard Methods adopted for examining water and wastewater.

Parameter	Standard method for the examination of water and wastewater
E. coli (MPN/100 mL)	9223B Enzyme substrate test with multi-well procedure
Total suspended solids (mg/L)	2540D Total suspended solids dried at 103–105 °C
Chemical oxygen demand (mg/L)	5220D Closed reflux, colorimetric method
Biochemical oxygen demand (mg/L)	SM 5210D Respirometric method
Total nitrogen (mg/L)	4500C Persulfate digestion method
Total phosphorus (mg/L)	4500-P E persulfate method
Potassium (mg/L)	4500B Potassium permanganate spectrophotometric method

before application in the constructed wetland (inflow) and in the outflow and the characteristics of the wastewater were determined by analyzing the following parameters: *E. coli*, BOD, COD, total phosphorus (TP), total nitrogen (TN), pH, potassium (K), and TSS (Table 2).

The constructed wetland system was efficient in removing solids and pollutants from the wastewater. In Phase 1 (50% wastewater +50% water), arithmetic mean reductions of 97.59% for *E. coli*, 8.21% for BOD, 44.53% for COD, 70.42% for TP, 32.73% for TN, 55.32% for K, and 65.23% for TSS were observed. For Phase 2 (75% wastewater +25%), arithmetic mean reductions of 92.69% for *E. coli*, 49.63% for BOD, 68.9% for COD, 66.66% for TP, 53.44% for TN, 37.81% for K, and 88.5% for TSS were recorded. In the Phase 3 (100% wastewater), the system was able to remove an arithmetic mean of 99.96% of the *E. coli*, 61.71% of the BOD, 80.05% of the COD, 83.99% of the TP, 73.35% of the TN, 54.56% of the K, and 95.94% of the TSS.

The constructed wetland system was able to treat the wastewater to the microbiological quality recommended for unrestricted irrigation (fecal coliform $\leq$1000 MPN/100 mL). In all three phases, the average concentration of *E. coli* in the wastewater influent to the system was 8.77×10^4 MPN/100 mL, while for the outflow, the mean concentration was 533.3 MPN/100 mL. The mean concentration of *E. coli* in treated wastewater was below the threshold recommended by the WHO for unrestricted reuse in agriculture.

7. Combination of biofiltration and SODIS

Biofiltration is one of the oldest wastewater treatment processes, it was first used in England in 1893 and since then, it has been successfully adopted in other regions worldwide.[14] The working principle of a biofiltration system involves the biodegradation of pollutants via microbial action rather than by physical straining, as occurs in a regular filter. A biofilter is any type of filter with biomass attached on the filter media. Several types of rocks, slag, and plastic media have been used as filter media; usually, the selection of the type of filter medium depends on local availability and the costs of transportation.

With the application of wastewater through a biofilter, microorganisms (e.g., fungi, bacteria, protozoa, and alga) will gradually develop as a biological film on the surface of the filter media; this is known as a biofilm. Biofilm formation may take a few days or even months, depending on the concentration of organic matter in the wastewater. The constant applications of organic substances and nutrients are essential to maintain microbial activity and to ensure that the performance of the biofilter remains consistent and effective.[14] Biofilters are efficient for treating wastewater with high organic loads, as they do not require large areas for installation, consume little energy, and are operationally simple. These traits make biofilters a good alternative for wastewater treatment in developing countries.

According to Jordão and Pessôa,[56] the pathogen reductions by biofiltration range from 90% to 95% for enteric viruses, 90–95% for bacteria, 50–90% for protozoa, and 50–95% for helminths, though, if the concentrations of pathogens in wastewater were high, biofiltration may not be enough to treat the wastewater to the microbiological quality level recommended for unrestricted irrigation. For instance, considering the concentration of fecal coliform in wastewater of 1.0×10^5 MPN/100 mL and assuming that the removal efficiency of this microorganism by the biofilter was 90%, the treated wastewater would have a fecal coliform concentration of 1.0×10^4 MPN/100 mL, which exceeds the threshold ($\leq$1000 MPN/100 mL) recommended by the WHO for unrestricted irrigation. For this reason, to achieve better microbiological quality, a disinfection method must be adopted. A suitable disinfection option for developing countries is SODIS.

A wastewater treatment system combining biofiltration and a concrete-based SODIS reactor was tested in Botucatu, São Paulo, Brazil by Silva.[57] The pilot study was conducted at the FCA and wastewater from the

Botucatu WWTP was used as the supply. The biofilter consisted of four rectangular fiberglass water tanks (100 × 35 × 31 cm) filled with gravel (size range: 4.8–9.5 mm) connected in series by 50-mm PVC pipes, with a difference of 2 cm between each water tank. The biofilter had a porosity of 65%, a useful volume of ~160 L, and a hydraulic retention time of approximately 1.06 days. During the study period, 150 L of wastewater was released through the biofilter. After passing through the biofilter, the wastewater was stored in a 2000 L plastic water storage tank. It was then directed to a concrete-based, black SODIS reactor shaped like an inverted truncated cone with a 1.00 m larger radius, 0.25 m smaller radius, and 0.30 m height (Fig. 10).

Fixed depths of wastewater (0.10, 0.15, and 0.20 m) were tested in the SODIS reactor (Fig. 11), considering different times of exposure to solar radiation. The behavior of the experimental system was monitored in two periods: from May to July of 2018 (autumn–winter), and from December of 2018 to March of 2019 (spring–summer). Twelve wastewater samples were collected (six in the autumn–winter period and six in the spring–summer period). Here, *E. coli* was adopted as the FIB to detect fecal contamination. In order to analyze the deactivation of *E. coli*, wastewater samples were collected every 2 h during a 10 h period of sunlight exposure (8:00 a.m. to 6:00 p.m., local time). The first wastewater sample was collected at 8:00 a.m. (0 h of sunlight exposure), the second at 10:00 a.m. (2 h of sunlight exposure), the third at 12:00 p.m. (4 h of sunlight exposure), the fourth at 2:00 p.m. (6 h of sunlight exposure), the fifth at 4:00 a.m. (8 h of sunlight exposure), and the sixth at 6:00 p.m. (10 h of sunlight exposure).

Global solar radiation data were obtained from the Lageado Weather Station (22°50′48.14″S, 48°25′53.52″W), which was located ~700 m from the experimental site. Mathematical models were then developed using the accumulated global radiation dose ($MJ\,m^{-2}$) and the ratio between the initial population (*No*) and the remaining population (N) of *E. coli* after exposure to solar radiation ($y = N/No$). The function used to analyze the data and adjust the models was an S-shaped function. In order to determine if the wastewater depth (0.10, 0.15, and 0.20 m) influenced the deactivation of *E. coli*, an analysis of variance (ANOVA) was performed. The efficiency of the biofilter was also calculated according to Eq. 3 for the following parameters: *E. coli*, BOD, COD, turbidity, and TSS. All wastewater samples were collected, preserved, and analyzed according to the *Standard Method for the Examination of Water and Wastewater* (Table 2).

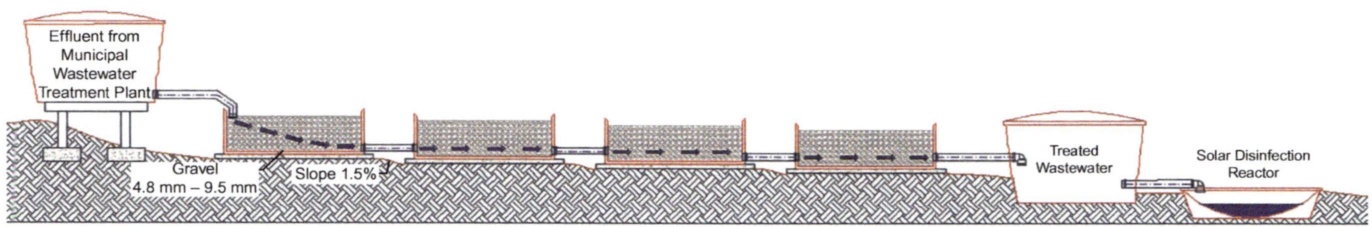

Fig. 10 Wastewater treatment system composed of a biofilter and SODIS reactor.

Fig. 11 Fixed depths (0.10, 0.15, and 0.20 m) of wastewater applied in the SODIS reactor.

$$E = \left(\frac{C_1 - C_2}{C_1}\right) \times 100, \quad (3)$$

In Eq. 3, E is the efficiency (%); C_1 is the concentration affluent (inflow), and C_2 is the concentration effluent (outflow).

The biofilter was efficient in removing organic matter and solids from the wastewater, presenting arithmetic mean efficiencies of 50.9% for BOD, 86.2% for COD, 95.8% for TSS, and 97.8% for turbidity. The wastewater from the Botucatu WWTP had mean values of 56.3 mg/L for BOD, 361 mg/L for COD, 397 mg/L for TSS, and 211 nephelometric turbidity units (NTUs). After passing through the biofilter, these means were reduced to 25.6 mg/L for BOD, 35.7 mg/L for COD, 10.7 mg/L for TSS, and 4.9 NTUs.

The wastewater pretreated by the biofilter and directed to the SODIS reactor reached the threshold established by the WHO for unrestricted irrigation ($\leq$1000 MPN/100 mL of fecal coliform) at 12:00 p.m., after 4 h of direct sunlight exposure, in both the autumn–winter and spring–summer study periods, independent of the wastewater depth applied (0.10, 0.15, or 0.20 m). The mean global solar radiation accumulated dose at 6 p.m. (10 h of sunlight exposure) was ~13.90 MJ m^{-2} in the autumn–winter period and 15.95 MJ m^{-2} in the spring–summer period. The initial mean concentration of *E. coli* in the wastewater in the SODIS reactor was $\geq 10^4$ MPN/100 mL, and after 10 h of sunlight exposure, it was reduced to $\leq$6 MPN/100 mL in the autumn–winter period and to $\leq$4 MPN/100 mL in spring–summer period. The low turbidity and concentration

Table 3 Simulated scenarios of the accumulated global solar radiation dose required for solar wastewater disinfection for reuse in unrestricted irrigation.

Initial concentration of *E. coli* MPN/100 mL (*No*)	Final concentration of *E. coli* MPN/100 mL (*N*)	Remaining population of *E. coli* (*N/No*)	Global accumulated dose of solar radiation (*RG*)
4.0×10^9	1000	0.00000025	$9.21\,MJ\,m^{-2}$
4.0×10^8	1000	0.0000025	$8.02\,MJ\,m^{-2}$
4.0×10^7	1000	0.000025	$6.84\,MJ\,m^{-2}$
4.0×10^6	1000	0.00025	$5.65\,MJ\,m^{-2}$
4.0×10^5	1000	0.0025	$4.52\,MJ\,m^{-2}$
4.0×10^4	1000	0.025	$3.35\,MJ\,m^{-2}$

of TSS contributed to the good performance of the SODIS reactor in the deactivation of fecal coliform (*E. coli*).

Since the ANOVA did not reveal any significant difference between the tested wastewater depths (0.10, 0.15, and 0.20 m), the data from the three wastewater depths for both periods were combined and adjusted into a single model, as shown in Eq. 4:

$$y = \frac{N}{No} = \frac{\exp\left(2.66149 - 1.93904RG\right)}{\left(1 + \exp\left(2.66149 - 1.193904RG\right)\right)}, \tag{4}$$

where RG is the accumulated global radiation ($MJ\,m^{-2}$) and y is the ratio between the initial population (No) and the remaining population (N) of *E. coli* after exposure to solar radiation.

The accumulated global solar radiation dose required for solar wastewater disinfection to the threshold recommended by the WHO for unrestricted irrigation ($\leq$1000 MPN/100 mL of fecal coliform) can be simulated by Eq. 4. Table 3 presents the results of some simulated scenarios.

8. Conclusions

The main purpose of wastewater reuse in developing countries is irrigation, mostly with untreated wastewater. The treatment of wastewater in many developing countries is affected by poorly developed legislation, a lack of regulatory mechanisms and policies to allow for gradual enhancement, a lack of data, technical issues, and lacking investment in wastewater collection and treatment infrastructure. Thus, low-cost wastewater treatment solutions, specifically SODIS reactors, constructed wetlands, and biofilters combined with SODIS reactors are viable solutions that may be adopted in developing countries for wastewater treatment in order to meet the quality standards recommended by the WHO for unrestricted irrigation, thus allowing its safe use as reclaimed water.

Acknowledgment

The authors would like to thank the São Paulo Research Foundation (FAPESP) for funding this work (grant nos. 2017/14729-3 and 2014/05997-6).

References

1. Wilas J, Draszawka-Bołzan B, Cyraniak E. Wastewater reuse. *World News Nat Sci* 2016;**5**(1):33–41.
2. United Nations World Water Assessment Programme. *Wastewater: The untapped resource.* Paris: United Nations World Water Assessment Programme; 2017.
3. United States Environmental Protection Agency. *Guidelines for water reuse [internet].* 2012: Washington: United States Environmental Protection Agency; 2012.1–4. Available from: https://www.epa.gov/sites/production/files/2019-08/documents/2012-guidelines-water-reuse.pdf.
4. Abdel-Dayem S, Taha F, Choukr-Allah R, Kfouri C, Chung C, Al SD. *Water reuse in the Arab world: From principle to practice—Voices from the field [internet].* Dubai: World Bank; 2012. Available from: http://documents.worldbank.org/curated/en/405461468136207446/Water-reuse-in-the-Arab-world-from-principle-to-practice-voices-from-the-field.
5. Martin-Hurtado R, Nolasco D. *Managing wastewater as a resource in Latin America and the Caribbean: Towards a circular economy approach [internet].* Available from: https://programme.worldwaterweek.org/Content/ProposalResources/allfile/managing_wastewater_as_a_resource_in_lac.pdf; 2020.
6. United Nations Economic and Social Commission for Asia and the Pacific (ESCAP), United Nations Human Settlements Programme (UN-Habitat) and Asian Institute of Technology (AIT). *Policy guidance manual on wastewater management with a special emphasis on decentralized wastewater treatment systems.* Bangkok: ESCAP, UN-Habitat, AIT; 2015.
7. *Wastewater management and sanitation in Asia and the Pacific [internet].* Asian Development Bank; 2011. [cited 1 June 2020]. Available from https://www.adb.org/features/promoting-wastewater-revolution-asia-adbs-plans-progress-and-initiatives.
8. Wang H, Wang T, Zhang B, Li F, Toure B, Omosa I, et al. Water and wastewater treatment in Africa—current practices and challenges. *Clean (Weinh)* 2013;**42**(8):1029–35.
9. Qadir M, Wichelns D, Raschid-Sally L, McCornick P, Drechsel P, Bahri A, et al. The challenges of wastewater irrigation in developing countries. *Agric Water Manag* 2010;**97**(4):561–8.
10. *African development fund. supplementary loan to Kampala Sanitation Program (AFKSP) [internet].* Available from: https://www.afdb.org/fileadmin/uploads/afdb/Documents/Project-and-Operations/Uganda_-_Kampala_Sanitation_Program__AFKSP_.pdf; 2017.
11. World Bank Group. *Water ppps in Africa [internet].* Available from: https://www.ifc.org/wps/wcm/connect/9a0e7b9a-f536-4232-9a6a-c3c60c1e8428/WBG_AfricaWaterPPPs.pdf?MOD=AJPERES&CVID=lKc4hl0; 2014.
12. Kivaisi A. The potential for constructed wetlands for wastewater treatment and reuse in developing countries: a review. *Ecol Eng* 2001;**16**(4):545–60.

13. Sanchez-Roman R, Soares A, Matos A, Sediyama G, Souza O, Mounteer A. Domestic wastewater disinfection using solar radiation for agricultural reuse. *Trans ASABE* 2007;**50**(1):65–71.
14. Chaudhary D, Vigneswaran S, Ngo H, Shim W, Moon H. Biofilter in water and wastewater treatment. *Korean J Chem Eng* 2003;**20**(6):1054–65.
15. Weldesilassie A, Boelee E, Drechsel P, Dabbert S. Wastewater use in crop production in peri-urban areas of Addis Ababa: impacts on health in farm households. *Environ Dev Econ* 2010;**16**(1):25–49.
16. Cornish GA, Kielen NC. Wastewater irrigation—hazard or lifeline? Empirical results from Nairobi, Kenya and Kumasi, Ghana. In: Scott C, Faruqui N, Raschid-Sally L, editors. *Wastewater use in irrigated agriculture: confronting the livelihood and environmental realities.* Wallingford, Oxfordshire, UK: CAB International in Association With the International Water Management Institute and the International Development Research Centre; 2004. p. 69–79.
17. Feenstra S, Hussain R, van der Hoek W. *Health risks of irrigation with untreated urban wastewater in the southern Punjab, Pakistan [internet].* Lahore: International Water Management Institute; 2000. Available from: https://publications.iwmi.org/pdf/H026997.pdf.
18. Contreras J, Meza R, Siebe C, Rodríguez-Dozal S, López-Vidal Y, Castillo-Rojas G, et al. Health risks from exposure to untreated wastewater used for irrigation in the Mezquital Valley, Mexico: a 25-year update. *Water Res* 2017;**123**:834–50.
19. Vollertsen J, Hvitved-Jacobsen T. Collection systems—dry and wet weather performance. In: Ujang Z, Henze M, editors. *Municipal wastewater management in developing countries: principles and engineering.* London, UK: IWA Publishing; 2006. p. 99–133.
20. Statistics Division of the United Nations Secretariat. *Methodology- standard country or area codes for statistical use (M49) [internet].* Unstats.un.org (web archive link, 01 June 2020) [cited 1 June 2020]. Available from: https://unstats.un.org/unsd/methodology/m49/.
21. *AQUASTAT database [internet].* Fao.org (web archive link, 01 June 2020); 2020 [cited 1 June 2020]. Available from: http://www.fao.org/nr/water/aquastat/data/query/index.html?lang=en.
22. Raschid-Sally L, Jayakody P. *Drivers and characteristics of wastewater agriculture in developing countries: results from a global assessment [internet].* Colombo, Sri Lanka: International Water Management Institute; 2008. Available from: http://www.iwmi.cgiar.org/Publications/IWMI_Research_Reports/PDF/PUB127/RR127.pdf.
23. Bazza M. *Wastewater reuse in the near east region: experience and issues [internet].* Iraklio, Crete, Greece: Food and Agriculture Organization of the United Nations Regional Office for the Near East; 2002. Available from: http://www.fao.org/3/ae528e/ae528e00.htm.
24. Food and Agriculture Organization of the United Nations—Regional Office for the Near East and World Health Organization Regional Office for the Eastern Mediterranean. *Expert consultation for launching the regional network on wastewater reuse in the near East [internet].* 2003. Available from: http://www.fao.org/3/a-y5186e.pdf.
25. Navarro I, Barrios J, Becerril E, Lucario S, Jimenez B. *Wastewater reuse for irrigation—practices, safe reuse and perspectives.* In: Javaid M, editor. *Irrigation and drainage—sustainable strategies and systems [internet].* ; 2015 [cited 2 March 2020]. Available from: https://www.intechopen.com/books/irrigation-and-drainage-sustainable-strategies-and-systems/wastewater-reuse-for-irrigation-practices-safe-reuse-and-perspectives.
26. Siebe C, Chapela-Lara M, Cayetano-Salazar M, Prado B, Siemens J. Effects of more than 100 years of irrigation with Mexico City's wastewater in the Mezquital Valley (Mexico). In: Hettiarachchi H, Ardakanian R, editors. *Safe use of wastewater in agriculture: good practice examples.* Dresden: UNU-FLORES; 2016. p. 121–38.
27. Uleimat A. *Wastewater production, treatment, and use in Jordan [internet].* [cited 1 June 2020]. Available from: https://jordankmportal.com/resources/wastewater-production-treatment-and-use-in-jordan; 2006.

28. Nazzal Y, Mansour M, Al Najjar M, McCornick P. *Wastewater reuse law and standards in the kingdom of Jordan [internet]*. Amman: The Ministry of Water and Irrigation; 2000. [cited 1 June 2020]. Available from: https://www.iwa-network.org/filemanager-uploads/WQ_Compendium/Database/Future_analysis/065.pdf.
29. Ammary B. Wastewater reuse in Jordan: present status and future plans. *Desalination* 2007;**211**(1–3):164–76.
30. Elmenoufy HM, Mostafa M, Eid MM, El Ganzoury A, El-Hussainy FM, Abdel Wahab MM. Towards enhancing rainfall projection using bias correction method: case study Egypt. *Int J Sci Res Sci Eng Technol* 2017;**3**(6):187–94.
31. Drechsel P, Hanjra MA. Wastewater for fruit and wood production (Egypt)—case study. In: Otoo M, Drechsel P, editors. *Resource recovery from waste: business models for energy, nutrient and water reuse in low- and middle-income countries*. Abingdon Oxon: Routledge; 2018. p. 556–68.
32. Dare A, Mohtar R, Jafvert C, Shomar B, Engel B, Boukchina R, et al. Opportunities and challenges for treated wastewater reuse in the West Bank, Tunisia, and Qatar. *Trans ASABE* 2017;**60**(5):1563–74.
33. Bahri A. Water reuse in Tunisia: stakes and prospects. In: Marlet S, Ruelle P, editors. *Atelier du PCSI (Programme Commun Systèmes Irrigués) sur une Maîtrise des Impacts Environnementaux de l'Irrigation*. 2002. p. 1–11. 2002 May 28–29; Montpellier, France: cirad-00180335.
34. Van der Hoek W, et al. *Urban wastewater: a valuable resource for agriculture a case study from Haroonabad, Pakistan*. Colombo: International Water Management Institute; 2002.
35. Singh P, Deshbhratar P, Ramteke D. Effects of sewage wastewater irrigation on soil properties, crop yield and environment. *Agric Water Manag* 2012;**103**:100–4.
36. Fink G, Alcamo J, Flörke M, Reder K. Phosphorus loadings to the world's largest lakes: sources and trends. *Global Biogeochem Cycles* 2018;**32**(4):617–34.
37. Wen Y, Schoups G, Van de Giesen N. Organic pollution of rivers: combined threats of urbanization, livestock farming and global climate change. *Sci Rep* 2017;**7**(1):1–9.
38. Angin I, Yaganoglu A, Turan M. Effects of long-term wastewater irrigation on soil properties. *J Sustain Agric* 2005;**26**(3):31–42.
39. Pescod M. *Wastewater treatment and use in agriculture—FAO irrigation and drainage paper 47*. 1992.
40. Carter L, Chefetz B, Abdeen Z, Boxall A. Emerging investigator series: towards a framework for establishing the impacts of pharmaceuticals in wastewater irrigation systems on agro-ecosystems and human health. *Environ Sci: Processes Impacts* 2019;**21**(4):605–22.
41. Martijn E, Redwood M. Wastewater irrigation in developing countries-limitations for farmers to adopt appropriate practices. *Irrig Drain* 2005;**54**(S1):S63–70.
42. Amoah I, Abubakari A, Stenström T, Abaidoo R, Seidu R. Contribution of wastewater irrigation to soil transmitted helminths infection among vegetable farmers in Kumasi, Ghana. *PLoS Negl Trop Dis* 2016;**10**(12):1–12.
43. Picó Y, Alvarez-Ruiz R, Alfarhan A, El-Sheikh M, Alobaid S, Barceló D. Uptake and accumulation of emerging contaminants in soil and plant treated with wastewater under real-world environmental conditions in the Al Hayer area (Saudi Arabia). *Sci Total Environ* 2019;**652**:562–72.
44. World Health Organization. *Guidelines for the safe use of wastewater, excreta and greywater*. World Health Organization; 2006.
45. Downes A, Blunt T. The influence of light upon the development of bacteria. *Nature* 1877;**16**(402):218.
46. Nelson K, Boehm A, Davies-Colley R, Dodd M, Kohn T, Linden K, et al. Sunlight-mediated inactivation of health-relevant microorganisms in water: a review of mechanisms and modeling approaches. *Environ Sci: Processes Impacts* 2018;**20**(8):1089–122.

47. *Disinfection by-products | the safe water system | CDC [internet]*. Cdc.gov (web archive link, 01 June 2020); 2020 [cited 1 June 2020]. Available from: https://www.cdc.gov/safewater/chlorination-byproducts.html.
48. Winward G, Avery L, Stephenson T, Jefferson B. Chlorine disinfection of grey water for reuse: effect of organics and particles. *Water Res* 2008;**42**(1–2):483–91.
49. García-Fernández I, Miralles-Cuevas S, Oller I, Malato S, Fernández-Ibáñez P, Polo-López M. Inactivation of E coli and E faecalis by solar photo-Fenton with EDDS complex at neutral pH in municipal wastewater effluents. *J Hazard Mater* 2019;**372**:85–93.
50. Queluz J, Sánchez-Román R. *Efficiency of domestic wastewater solar disinfection in reactors with different colors*. *Water Util J* 2014;**7**(1):35–44. Available from: http://ewra.net/wuj/pdf/WUJ_2014_07_04.pdf.
51. Queluz JGT. *Estudo da eficiência da desinfecção solar de águas residuárias domésticas em reatores de diferentes colorações*. Botucatu, SP: School of Agronomic Sciences-FCA, Unesp; 2013.
52. APHA, AWWA and WEF. *Standard methods for examination of water and wastewater*. 22nd ed. Washington: American Public Health Association; 2012.
53. Herath I, Vithanage M. Phytoremediation in constructed wetlands. In: Ansari A, Gill S, Gill R, Lanza G, Newman L, editors. *Phytoremediation management of environmental contaminants*. vol. 2. Springer International Publishing; 2015. p. 243–63.
54. Queluz JGT. *Efficiency of constructed wetlands for low strength domestic wastewater treatment [PhD]*. Botucatu, SP: School of Agronomic Sciences-FCA, Unesp; 2016.
55. Cunha AR, Martins D. Climatic Classification for the Districts of Botucatu and São Manuel, SP. *Irriga* 2009;**14**(1):1–11.
56. Jordão E, Pessôa C. *Tratamento de esgotos domésticos*. 6th ed. Rio de Janeiro: ABES; 2011.
57. Silva TLS. *Efficiency analysis of a low-cost wastewater treatment system for agricultural reuse in rural communities and traditional populations [master]*. Botucatu, SP: School of Agronomic Sciences-FCA, Unesp; 2019.

CHAPTER TWO

Water reuse in India: Current perspective and future potential

Arun Kumar* and Kirti Goyal
Department of Hydro and Renewable Energy, Indian Institute of Technology, Roorkee, India
*Corresponding author: e-mail address: arun.kumar@hre.iitr.ac.in

Contents

1. Introduction 34
   1.1 Water supply-demand gap 35
   1.2 Driving factors for reclaimed water reuse 38
   1.3 Challenges encountered in reclaimed water reuse 39
2. Potential for reclaimed water reuse in India 40
   2.1 Additional source of water supply 41
   2.2 Resource recovery options 41
   2.3 Employment and revenue generation 41
   2.4 Energy savings 43
3. Applications of treated wastewater in India 45
   3.1 Agricultural reuse 45
   3.2 Urban reuse 47
   3.3 Environmental/recreational reuse 48
   3.4 Industrial reuse 48
   3.5 Indirect and direct potable reuse 48
4. Opportunities for wastewater reuse in India 49
   4.1 Technical feasibility 50
   4.2 Financial viability 52
   4.3 Legal initiatives 55
   4.4 Market viability 59
5. Conclusions 60
References 61

Abstract

The rising water scarcity throughout the world has stimulated the use of treated wastewater as an alternative option. Wastewater is a valuable resource that can be used beneficially to meet the water supply-demand gaps. In India, water and wastewater sector, especially wastewater reuse, has recently started getting attention due to growing water problems specially in water scarce cities. This chapter presents an overview of wastewater reclamation and reuse in India with the focus on its rationale, current practices and challenges, potential and future scope. Urban local bodies and municipalities

Advances in Chemical Pollution, Environmental Management and Protection, Volume 6
ISSN 2468-9289
https://doi.org/10.1016/bs.apmp.2020.07.011

have implemented reclaimed water reuse projects in many water sensitive cities in the country. However, the challenges such as lack of planning and feasibility studies, incentives, social acceptance, regulatory guidelines and poor institutional framework have made it difficult to sustain such projects. There is a large potential for treated wastewater reuse in the country owing to high quantities of wastewater. Studies show that if 80% of urban wastewater could be treated by 2030, there would be an increase of 400% in the volume of available wastewater to reclaim and directly reuse. In view of the rising opportunities, the reuse in the industrial and agricultural sectors has grown in the past decades but in the municipal sector, it is still evolving. Many legal initiatives have been taken by the central and state governments over the past few years. Research shows that to establish widespread and successful reclaimed water reuse, an integrated water resources planning and management approach is imperative.

Keywords: Integrated approach, Market viability, Reuse potential, Technical feasibility, Treated wastewater, Water reclamation, Wastewater reuse, Water scarcity, Water stress

1. Introduction

Water is the main source of existence of the living beings. It is well known across the global organizations that water plays a substantial role in achieving sustainability. The civilization needs water for majority of its activities such as agriculture, industries and domestic purposes, etc. However, access to safe freshwater has become quite challenging nowadays due to many factors. Throughout the world, the availability of water is threatened owing to population growth, climate variations and changes in the land and water use patterns. Further, human activities have laid a negative impact on the ecological cycle of water such as the reduction of forest areas and grasslands, urbanization leading to reduced groundwater recharge, overexploitation of groundwater resources, seawater intrusion, pollution of surface water resources due to untreated industrial and municipal discharge, etc., adversely affecting the water quantity and quality.

According to the United Nations Environment Programme (UNEP), water deficiency and global warming are the two most disturbing problems worldwide. In the last century, the global water use has escalated by a factor of six and is gradually increasing at a rate of about 1% per year. Within the next 50 years, the world population is expected to increase by 40–50%.[1] The population growth, coupled with industrialization and urbanization, will result in a highly increased demand for water.

To address the existing and anticipated water shortages, water resource planners are seeking alternative sources to enhance the limited freshwater resources at a global level by the use of recycled or reclaimed water, most commonly known as, reuse of treated wastewater. It is defined as the use of water which is generated from wastewater and that achieves, after treatment as necessary, a quality that is appropriate (taking in account the health and environment risks) for its intended use such as irrigation, industrial or civil (municipal) in general. Reuse of treated wastewater has emerged as an environmentally sustainable option that has helped many global communities significantly in augmenting their water supplies.

This chapter focusses on discussing the existing scenario of water reuse in India wherein, the main driving factors and challenges related to reclaimed water reuse are described. An elaborative view of the potential of the treated wastewater reuse in terms of benefits, technical feasibility and economic viability, future opportunities with the help of several case studies has been presented.

1.1 Water supply-demand gap

India is imminent to be the most populated country in the world which in turn will have higher water demands. The geographical area of India covers approximately 329 million hectares (2.45% of the earth's landmass) containing 4% of the world water resources.[2] Though the average annual water resources potential is 1869 Billion Cubic Meters (BCM) (Table 1), the amount of water that can be actually used beneficially is much less due to severe limitations posed by physiography, topography, political issues and the present state of technology to harness water resources economically. India receives an annual rainfall of approximately 4000 BCM which is more than enough to fulfill the water demand. However, only a small portion of this rainfall could be used effectively owing to its variability over time and space, limiting the water availability and available surface and groundwater storage.[3]

Table 1 Water availability in India.[2]

S. No.	Availability	Quantity (BCM)
1	Average Annual Precipitation (including snowfall)	4000
2	Natural Runoff (Surface Water and Groundwater)	1869
3	Estimated Utilizable Water Resources	1123
	(i) Surface Water Resources	690
	(ii) Ground Water Resources	433

The water problems in India mainly comprise of spatio-temporal variation of water resources leading to lesser water availability even for the present population, exhausting surface and groundwater resources and highly polluted water bodies. There are 351 polluted river stretches in the country with 45 of them being critically polluted.[4] It has been estimated that by the year 2030, the anticipated water demand would reach 1498 BCM against an expected supply of 744 BCM which will widen the gap between demand and supply.[5](Fig. 1)

The country is close to reaching its water resources limits with more than 54% of the covered area facing high to extreme water stress (<1700 m^3/year/person of water availability). The existing water resources are under significant pressure due to high water demand patterns (Table 2). India has surpassed China to become the largest source of energy related water demand with an expected water withdrawal of 68 BCM by the year 2040. Based on utilizable resources, according to 2011 census, for a population of 1.2 Billion, the country has only 1000 m^3/year/person of water availability (Fig. 2), indicating the onset of severe water scarcity in many regions.[6]

The current water situation clearly indicates that there is a substantial gap between the water supply and demand and it is expected to amplify with growing population and other factors. To deal with this water crisis, the Government of India has attempted many efforts at state and central level such as, diversion of river flows to upsurge the surface water resources

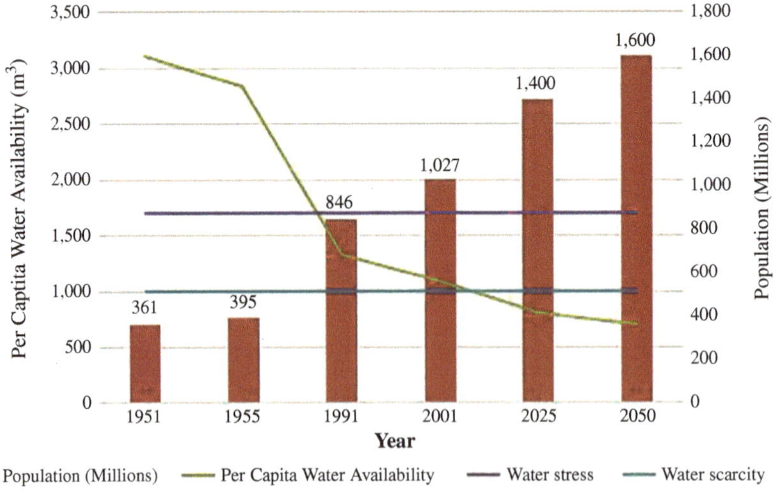

Fig. 1 Projected per capita water availability w.r.t Population growth.[2]

Table 2 Projected water demand in India.[2]
Water demand in BCM

Sector	2010	2025	2050
Irrigation	688	910	1072
Drinking Water	56	73	102
Industry	12	23	63
Energy	5	15	130
Other	52	72	80
Total	813	1093	1447

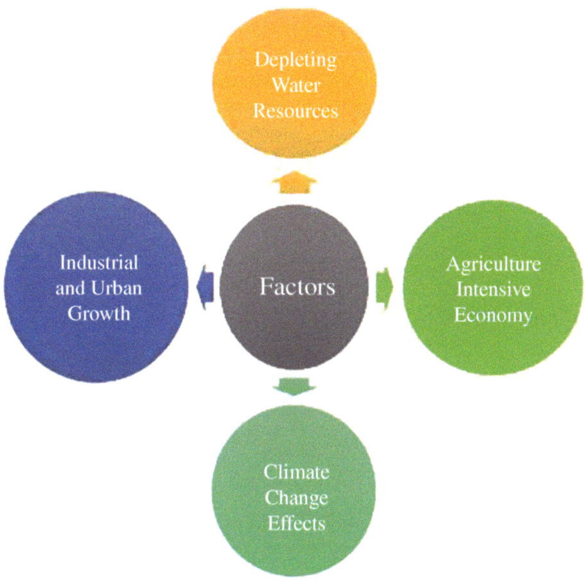

Fig. 2 Major factors driving reclaimed water reuse.

and water conservative agricultural practices, etc. One such pathway to alternative water supply is the reuse of treated wastewater.

The concept of reuse is not new to India. The first instance of wastewater reuse in India goes back to 1964–65.[7] It was observed that about 20% of the wastewater generated from the textile industry could be reused without any pre-treatment for certain applications like blanket washing, thereby, reducing the costs.[7] Also, the Air India building in Mumbai was the first commercial building in India to treat blackwater generated from toilets and reuse it as cooling water in centralized air conditioning systems.[8] The use of wastewater for irrigation dates back to the 1970s in Chennai. Nowadays, there are many operational recycling and reuse projects in several cities, such as Nagpur, Vadodara, Kolhapur, Chandigarh, Ahmedabad, Tuticorin, Vishakhapatnam, Kolkata, Nashik and Aurangabad, etc.[9] Lately, India has emerged as one of the fastest growing markets for reuse of treated wastewater, especially for industrial reuse owing to various factors as discussed further.

1.2 Driving factors for reclaimed water reuse

Rising Population, changes in rainfall patterns, decline in water availability, industrial development and rapid urbanization have together put the issue of water availability at the utmost priority for the Government. Factors such as water scarcity due to the increase in per capita consumption, technological development, environmental and economic challenges, etc., necessitate the reuse of treated wastewater.

1.2.1 Agriculture intensive economy

Being an agro-economic country, high amounts of food have to be produced leading extensive water demand. The demand for food is expected to reach 450 Megatons for 1600 Million people by the year 2050.[2] Moreover, the dependence on groundwater for agriculture is also quite high which has led to the search for augmenting the existing water sources.

1.2.2 Depleting water resources

Surface water is drying up due to anthropogenic activities and extreme weather conditions. 275 out of 445 rivers and 76 out of 85 lakes are highly polluted, rendering the water unfit for use.[10] Groundwater has declined by 61% in the last decade with net annual withdrawal being more than the recharge.[11] Major Indian cities like New Delhi, Hyderabad and Chennai, etc., are expected to run out of groundwater by the end of year 2020.[5]

1.2.3 Climatic variation effects

67% of the Himalayan mountains have retreated due to global warming causing lesser water in the streams.[5] Further, there is a reduction in the number of rainy days annually.[12] The society is vulnerable to the effects of climate change which has hampered the water resources planning and management.

1.2.4 Industrialization and urbanization

Being a developing economy, higher industrial output is expected which is directly linked to high water demand. Further, the urban population is increasing rapidly. By 2030, five of the world's largest cities will be in India, impacting the water resources deeply.[6] Urbanization, together with agriculture and industrialization are the major socio-economic factors affecting the water supplies.

1.3 Challenges encountered in reclaimed water reuse

India is still in the initial phases of wastewater reclamation and reuse. There are many barriers associated with the implementation of treated wastewater reuse projects, limiting the growth of wastewater reclamation and reuse in the country.

1.3.1 Technical challenges

Due to low sewerage network coverage (one-third of the cities have sewerage systems), insufficient/sub-optimal Sewage Treatment Plant (STP) capacity, obsolete and faulty sewer networks, the volume of wastewater available for reuse is quite limited.[4] There is a wide gap between the amount of wastewater generated and treated. Only 30% of the municipal wastewater and 60% of the industrial wastewater is treated.[4] Also, there is a limitation of reliable treatment technologies for removal of required contaminants for appropriate reuse and availability of skilled personnel having the required knowledge of the functioning of wastewater reuse treatment units.

1.3.2 Institutional challenges

There is no clear and must policy setting regarding the reuse of treated wastewater. Lack of institutional and regulatory structure obstructs the establishment of the treated wastewater reuse systems. Water, being a state subject, there is a lack of coordination between various sectors responsible for water supply and wastewater management. This fragmentation leads to delays in planning and implementation, unnecessary disagreements, division of resources and complexity in interdepartmental agreements which results in increasing the complexity of reclaimed water reuse projects more than required. Due to fragmentation of water sector, there is a limitation of data availability regarding water supply-demand gaps, wastewater volumes and water consumption, etc., which impacts the reclaimed water reuse planning. Surplus availability of water in smaller cities and towns has made the local bodies highly dependent on the available resources.

1.3.3 Economic challenges

The initial costs for new treated wastewater reuse programs are too high and the access to finance and technology for the establishment of such projects is quite limited. The long-term economic viability of reuse projects is crucial for effective reclaimed water reuse. In most of the cities, the water charges for domestic and industrial users vary with the industrial tariff being significantly higher than the domestic tariff. In those cases, the income generated from industrial charges is used to compensate for the municipal/domestic water supply costs. By reusing treated wastewater, a part of this additional revenue could be eliminated which creates unwillingness among the local administrative units for reuse of treated wastewater.

1.3.4 Social challenges

No attention is paid to the public awareness and perception which leads to the failure of reclaimed water reuse projects due to lack of community trust in government authorities, especially for water quality and health related issues. Inadequate treatment technologies and associated public health and risks further drive the public fear to go for the reuse of treated wastewater. The perceptions of public utilities regarding consumers' reactions on reuse of reclaimed water poses as a barrier more than the reaction itself.

These conditions hold back the development of treated wastewater reuse. But, due to development in technology and economic growth, the situation is fast changing. There is a high potential for reuse of reclaimed water in the country and lately, many opportunities for the same have been taken up as presented later.

2. Potential for reclaimed water reuse in India

On the verge of being one of the severely water scarce countries in the world, due to its high population, there is a lot of potential for reuse of treated wastewater in the country. As India is an agro-economic and developing nation, its water resources, especially, groundwater, is under stress to supply water for both agricultural and industrial activities. To ensure a sustainable supply of water for such purposes, reclaimed water plays a major role in the Indian context. Recycling and reuse of treated wastewater is quite favorable as it provides an opportunity for environmental restoration by keeping the wastewater discharge out of water bodies along with providing an assured supply of water to meet the increased water demands, thereby, reducing the pressure on freshwater resources.

According to a study, by the year 2030, a collection of 80% of urban wastewater would generate a total volume of approximately 17 BCM water annually. Adequate treatment and subsequent reuse of this volume could fulfill 75% of the industrial water demand and almost one-fourth of potable water demand. The collection and treatment of 38,000 Million liters per day (MLD) of urban wastewater every year could provide 14 BCM water, which could be directly reused to irrigate up to 3 Million hectares (Mha) area. The study shows that recycled wastewater from 465 class I (100,000 and above population) and 3894 class II cities (population between 50,000 and 99,999) has the potential to meet one-fourth of the existing industrial water demand (17 BCM including energy demand).[13]

2.1 Additional source of water supply

Reclaimed water ensures a secure source of water supply and aids in augmentation of existing water supplies. It is cheaper than other options like desalination. Because of volumetric tariff in many cities like Chennai, Hyderabad, Bengaluru and Mumbai, etc., treated water is less-expensive in comparison to fresh water supply. Due to limited groundwater availability in India, treated water seems to be a reliable and preferable option for agriculture and especially for industries, as they have higher charges than municipal tariffs. Moreover, industries are generally situated at the outskirts of the cities, thereby, reducing the conveyance cost of treated wastewater. It will also lead to savings on freshwater abstraction costs. Reclaimed water is used to fulfill 15% of total water demand in Chennai city and 10% of total water demand in Bengaluru city (Box 1).[13,14]

2.2 Resource recovery options

Reuse of treated wastewater provides opportunities to recover multiple nutrients like nitrogen, phosphorus and potassium, etc., as well. The sludge generated in the treatment process has high nutritional value and is used as a fertilizer/manure in many areas. India imports most of its potassium and phosphorus. Recovering these nutrients from treated wastewater may help in reducing the import costs (Box 2).

2.3 Employment and revenue generation

Reuse of wastewater has many positive implications, such as the employment generation and revenue generation. One way to recover the capital as well as operation and maintenance costs of a water reuse project is to sell

BOX 1 There are many studies which have reported the reuse of treated wastewater to supplement the water supply.

- As per a study conducted in Hyderabad, for an anticipated population growth rate of 2.5%, collection and use of 90 Million cubic meters (MCM) of runoff, adoption of water harvesting by 0.5 million households and reuse of 120 MCM of treated wastewater annually would fulfill the city's water by the year 2031.[15]
- Another research shows that out of the total amount of 794 MLD supplied to residential areas in the Pune city, 42% of the demand could be met by reusing treated wastewater.[16]
- Researchers assessed the urban water security in the National Capital Territory of Delhi and observed that there is a reuse potential of 78.2% if proper collection and management of the generated wastewater takes place.[17]
- IIT Madras has a water demand of 2.8 MLD whereas, the Chennai Metropolitan Water Supply and Sewerage Board is able to supply 1.2–1.8 MLD of water. The institute has bridged the gap by reusing 1.4 MLD of treated wastewater for toilet flushing and gardening.[18]

BOX 2 Many case studies have reported that reuse of treated wastewater has many resource recovery options as discussed.

- Usage of wastewater for irrigation along the Musi river in Hyderabad has helped the farmers in reducing the fertilizer costs and the reliable supply has aided in increasing the cropping intensity.[19]
- In Jaipur, nutrient recovery from sludge in the form of manure is used for agriculture and nursery purposes.[13]

the treated water to industries subject to the desired treatment levels. On an average, the use of treated wastewater for irrigation has resulted in a 30–40% increase in crop yield, more than 50% decrease in fertilizer use, a 30% increase in farmer's income and an average of INR 17,000/ha of economic benefits in many cities.[9] Apart from the monetary benefits, recycling and reuse has non-monetary benefits as well. Treated wastewater reuse also helps in the ecological remediation of water bodies by reducing the pollution and maintaining aquatic biodiversity (Box 3).

2.4 Energy savings

Reuse in irrigation can help in reducing groundwater pumping, decreasing the energy requirements which in turn reduces greenhouse gases.[13] This may provide opportunities to gain carbon credits for additional revenue generation. Studies show that it has a potential to reduce 1.75 Million MWh of electricity, which in turn reduces 1.5 Million tonnes of CO_2 GHG emissions annually (Box 4).[13,22]

A circular water cycle is preferred over a linear one because of its numerous benefits as shown in Fig. 3.

BOX 3 Reclaimed water reuse offers several employment and revenue generation options as well.

- In Hyderabad, cultivation of Jasmine flowers using wastewater has provided employment to approximately 120 farmers. The farmers can earn upto INR 20,000 per hectare for a 9 month flowering season.[20] Also, the use of tertiary treated sewage in the Bengaluru city generates a monthly revenue of about INR 4 Million along with augmenting the water supply.[13]
- The East Kolkata sewage fisheries is the largest reclaimed water reuse aquaculture system in the world. The wetland aids in cultivation of 5100 ha of land and supports 100,000 direct stakeholders. It also provides employment to 70,000 people annually.[21]

BOX 4 Reuse of treated wastewater is an eco-friendly and economic option.

Renaissance Hotel and Convention Centre in Mumbai had been facing water shortage issues, especially in operating the AC cooling towers for three very large high-rise hotel buildings. To deal with the same, they upgraded their STPs to accommodate recycling and reuse and installed a dual piping system along with water conservation measures. The use of recycled water for flushing toilets has reduced the intake consumption by 45% which has resulted in huge water bill savings.[23]

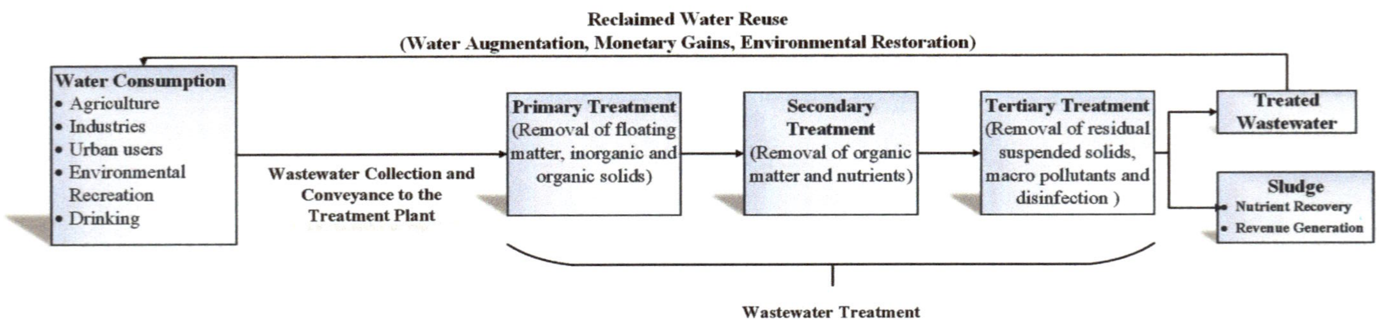

Fig. 3 A Schematic diagram showing the benefits of a closed loop water cycle.

3. Applications of treated wastewater in India

Depending on factors like wastewater quantity and quality, public awareness, market feasibility, type of treatment and distribution system, cost incurred, etc., there are a number of feasible options for the reuse of treated wastewater. The practice of reclaimed water reuse may be grouped into five main categories as shown in Fig. 4.

3.1 Agricultural reuse

The largest percentage of wastewater reuse (both treated and untreated wastewater) throughout the world mainly consists of agricultural reuse and it has become quite popular especially in developing economies with high population, like, India. Agricultural reuse using sewage is in practice for ages. It includes irrigation of both food as well as non-food crops. Untreated wastewater has been used in India for quite some time leading to many negative implications such as public health risks due to presence of pathogens, increased soil salinity, eutrophication and heavy metal accumulation. In Musi River basin, Hyderabad, reuse of wastewater for irrigation resulted in reduction of rice production by 40–50% due to presence of high salt levels.[24] Also, in Varanasi city, the ministry of Agriculture has recently notified for not using the wastewater in the city even for agricultural purposes due to the presence of metals in treated wastewater because of the

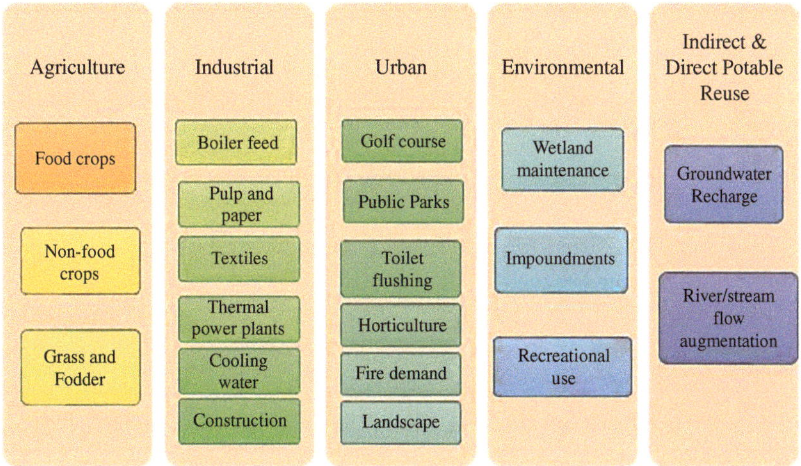

Fig. 4 Various categories of reclaimed water reuse.

intermixing of sewage with effluent from small scale home-based textile industries.[25] The studies clearly indicate that the source water quality plays a major role in planning for reclaimed water reuse.

Thus, secondary treatment along with disinfection is the minimum requirement for agricultural reuse which may vary in different regions depending on the effluent quality and available technologies. The main water quality parameter of concern is Fecal coliform. The Central Public Health Engineering and Environmental Organization (CPHEEO) Manual on Sewerage and Sewage Treatment (2013) enlists a detailed discussion on agricultural reuse in India and has suggested certain minimum water quality standards which must be followed (Table 3) (Box 5).

Table 3 Water quality guidelines for treated water reuse applications.[26]

Parameter	Units	Urban use	Agricultural use	Environmental use
pH	6.5–8.3			
BOD_5	mg/L	≤10	≤10	≤10
Turbidity	NTU	≤2	Nil	≤2
TSS	mg/L	Nil	Nil (for crop eaten raw) & <30 (for crops eaten cooked or non-edible crops)	Nil
Fecal coliforms	No./100 mL	Nil	Nil (for crop eaten raw) & <230 (for crops eaten cooked or non-edible crops)	Nil
Chlorine residual	mg/L	≥1	Nil	≥0.5
Intestinal nematodes	No./liter	NIL	<1	Nil
Total Kjeldahl Nitrogen (as N)	mg/L	<10	<10	<10
Dissolved Phosphorus (as P)	mg/L	<1	<2 (for crop eaten raw) & <5 mL (for crops eaten cooked or non-edible crops)	1

3.2 Urban reuse

Urban reuse has recently gained popularity in India, especially in the cities. Urban reuse can be unrestricted as well restricted. Turbidity, fecal coliform and residual chlorine are the concerning water quality parameters for urban reuse which makes it mandatory for wastewater to receive significant tertiary treatment and advanced disinfection so as to not be considered as a threat to the public health. Suggested minimum water quality criteria for urban non-potable reclaimed water reuse are given in Table 3 (Box 6).

BOX 5 Treated wastewater is extensively used for irrigation in many parts of the country.

- In New Delhi, near Keshopur and Okhla STP 1700 ha of land is irrigated using treated wastewater to grow vegetables. In Ahmedabad and Kanpur, wheat crop is widely irrigated with treated wastewater.[27]
- In Hyderabad, 2100 ha of land along the Musi River is irrigated with treated wastewater to cultivate paddy crop. Treated wastewater is also used for growing vegetables throughout the year. Moreover, about 10,000 ha of land is irrigated to grow para grass, a kind of fodder. Horticulture of public parks and trees is also done using treated wastewater.[28]
- Farmers in Kanpur and Hyderabad grow flowers with wastewater.

BOX 6 Urban reclaimed water reuse mainly for toilet flushing is done in many urban cities.

- In Pune city, nearly 567 MLD of treated wastewater is reused for industrial and non-contact domestic activities like toilet flushing, house cleaning, etc.[16]
- In Delhi, approximately 400 MLD of treated wastewater is reused throughout the city for horticulture, non-contact domestic activities and industrial uses, especially in construction activities.[29]
- The effluent generated from the wastewater treatment plants at the Chhatrapati Shivaji Maharaj International Airport (Mumbai), the Indira Gandhi International Airport (New Delhi) and the Rajiv Gandhi International Airport (Hyderabad) is treated and used for toilet flushing, greenery and as make up grade water for Heating, Ventilation and Air conditioning (HVAC) which has resulted in reduction in the water demand.[18,30]

3.3 Environmental/recreational reuse

In India, environmental reuse encompasses the use of treated wastewater to support wetlands, rejuvenation of existing lakes, creating artificial lakes, recreational purposes to enhance tourism, livestock maintenance, etc. One such example is the use of treated wastewater to supplement evaporation losses in the Hussain Sagar lake in Hyderabad.[8] Though, the scale of environmental reuse is limited as compared to other uses, it is expected to gain more attention in the coming future. Thus, certain water quality requirements need to be maintained for safe environmental reuse, thereby, making the water safe against health and environmental risks (Table 3).

3.4 Industrial reuse

In India, the industries are more susceptible to the rising water scarcity as they are given the last preference by the National Water Policy (NWP) during water allocation. In addition to that, high water footprint of Indian industry has escalated the reuse of treated wastewater for industrial purposes throughout the country. Use of reclaimed water for secondary activities in or out of the industry is being practiced successfully in many regions. Reclaimed water has high scope in water intensive industries like thermal power plants, pulp and paper, textiles, tanneries, oil/gas, pharmaceuticals, mining and metal processing, etc. Construction industry has recently gained high popularity in many urban cities in India. In the last few years, these industries have used treated wastewater for various purposes ranging from process water, boiler feed water, in cooling towers, toilet flushing and irrigation at site (Box 7).

For industrial reuse, tertiary treatment including disinfection along with nitrogen and phosphorus control is a must. Depending on the industrial applications, the water quality requirements also vary as shown in Table 4.

3.5 Indirect and direct potable reuse

Water reclamation for non-potable applications is still in its early stages in India. The reuse of treated wastewater to augment potable water supplies to meet the impending water demands is quite significant. It requires a high level of tertiary treatment using micro-filtration with reverse osmosis, nitrogen and phosphorus removal and disinfection and high level quality monitoring. As of now, there are no guidelines available for potable reuse in India. Though, de-facto reuse of treated wastewater is being done in the country for a long time due to disposal of wastewater in the rivers and

BOX 7 Treated wastewater has wide applications in Indian industries.

- Chennai Power Corporation Limited (CPCL) has treated 2.5 Million Gallons per day (MGD) by tertiary treatment and reverse osmosis for reuse of treated wastewater. CPCL is also executing an additional 2.5 MGD sewage recycling and reuse in the city plant along with a 5.8 MGD sea water desalination plant.[31]
- In Chennai, Madras refineries produce 12 MLD of reusable water and Madras Fertilizers Ltd. (MFL) produce 16 MLD of reusable water.[26]
- The Rashtriya Chemicals and Fertilizers (RCF) industry in Mumbai uses 23 MLD treated water as cooling grade makeup water.[32]
- Pragati Power Corporation Limited (PPCL), New Delhi use over 90% of their treated wastewater for cooling requirements.[31]
- The Bengaluru Water Supply and Sewerage Board (BWSSB) supplies tertiary treated wastewater to nearby industries to meet the non-potable water requirements in the city. Treated wastewater from the 10 MLD Yelahanka Tertiary treatment plant is supplied to the International Airport, Indian Tobacco Company, Rail Wheel Factory, Bharath Electronic Limited and the Indian Air Force. The treated effluent from the 180 MLD Vrishabhavathi Valley wastewater treatment plant is also supplied to a number of industries.[18]
- The Gurugram District Authority has made it compulsory for all the construction firms to reuse treated wastewater generated from the sewage treatment plants for the construction and other non-potable purposes.[33]
- The Surat Municipal Corporation (SMC) is currently supplying treated wastewater from Bamroli STP to industrial units in the Pandesara Industrial Estate for non-potable uses.[34]
- In Vadodara, tertiary treated industrial wastewater is done is supplies for non-potable industrial purposes throughout the city. Around 75% of the treated wastewater is available for reuse at a treatment cost of INR 36/1000 L.[9]

groundwater injection. However, planned potable water reuse is an area of future growth which needs a lot of planning and research before its promotion and application.

4. Opportunities for wastewater reuse in India

Though wastewater reclamation and reuse is in a nascent stage in India, over the past few years, the focus has been shifting to find solutions for the challenges encountered and promote the reuse of treated wastewater

Table 4 Water quality guidelines for treated water reuse applications in the industries.[26]

	Industrial application				
Constituent, mg/L	**Boiler feed**	**Pulp and paper**	**Textiles**	**Petroleum and coal**	**Cooling water**
Calcium	0.01–0.4	20	–	75	100
Iron	0.05–1.0	0.3–1.0	0.1–0.3	1	–
Manganese	0.01–0.3	0.05–0.5	0.01–0.05	–	–
Alkalinity as CaCO3	40–350	100	–	125	
Chloride	–	200–1000	–	300	100
TDS	200–700	–	100	1000	–
Hardness as CaCO3	0.07–350	100	25	350	–
Ammonium-N	0.1	–	–	–	1–3
Phosphate-P	–	–	–	–	0.6
Silica	0.7–30	50	–	–	20
Color (Hazen)	–	10–30	5	–	–

in the nation. There are various existing and upcoming opportunities in water and wastewater sectors leading to success of treated wastewater reuse.

4.1 Technical feasibility

Advancements in treatment processes and other technological innovations have enhanced the quality of effluent available for reuse. The selection of a specific treatment train for reclaimed water reuse depends on many geographical, demographical and financial conditions. Generally, treatment techniques are categorized in two main groups, i.e., location and time (Fig. 5).

Selection of a treatment train depends highly on the funding options and the potential use of the reclaimed water. The level of treatment required is directly linked to the cost and its use (Fig. 6). Higher the quality of water required, higher is the treatment costs.

Conventional wastewater treatment involves a combination of physical, chemical, and biological processes and operations to remove inorganic solids, organic matter, and nutrients like nitrogen and phosphorus from the wastewater. There are many conventional technologies available for

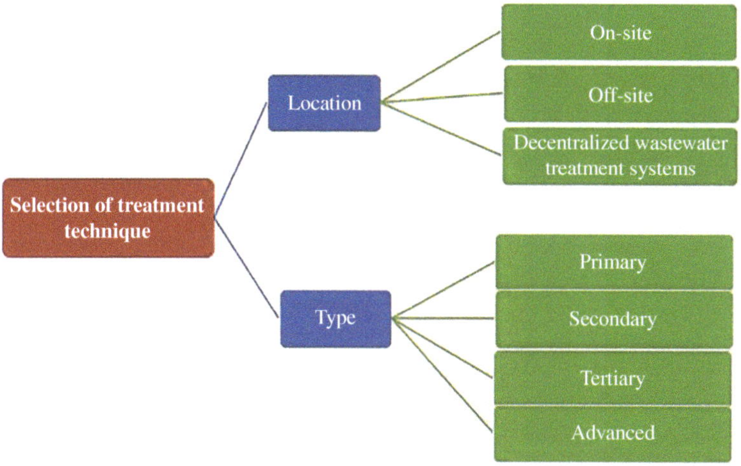

Fig. 5 Categorization of treatment techniques available for water reuse.

Fig. 6 Inter-linkage of various factors in selection of treatment process.

treatment of wastewater for reuse. A number of the most commonly used processes include activated sludge process and its modifications like sequential batch reactor, extended aeration, etc., upflow anaerobic sludge blanket, fluidized aerobic bed reactor, etc. Research progression has led to development of high level treatment technologies such as membrane bioreactor which can treat the wastewater almost up to the quality of river water.

Other than conventional treatment systems, natural systems such as waste stabilization ponds, constructed wetlands, have also gained attention in the recent times. Natural technologies have advantages over conventional processes in terms of lesser costs, low energy and labor requirements, etc. Lately, decentralized wastewater treatment technologies (DWWT) have emerged, especially for low income, highly populated countries like India as they are affordable and require lesser infrastructure and energy unlike

BOX 8 Some of the decentralized wastewater treatment reuse case studies have been discussed here.

Naval Civilian Housing Colony—Kanjurmarg, Mumbai[35]

The DWWT system employing Soil Biotechnology (SBT) is treating 50 m^3 per day wastewater generated from 168 households and use it for meeting the horticultural needs.

National Environmental Engineering Research Institute (NEERI), Worli, Mumbai[35]

NEERI is treating 5 m^3 per day of graywater generated from the bathrooms, kitchens and laboratories using the Phytorid technology. The treated wastewater is locally used for meeting horticultural needs.

Sangli City, Maharashtra[35]

The city has installed an individual household building DWWT system. 1 m^3 of the wastewater is treated using the Fixed Film Biofilter technology (FFBT). The treated wastewater is used for meeting the horticultural demand.

Auroville, Puducherry[13]

The Sangamam Housing Project implemented on the outskirts of Auroville, a township situated amidst the state of Tamil Nadu and the union territory of Puducherry has a DWWT system wherein, an anaerobic up flow reactor is used for primary treatment, a Root Zone secondary treatment system followed by maturation ponds. The treated wastewater is used for toilet flushing and gardening, etc. There is a reported reduction of 120 L per capita per day (lpcd) in the demand for freshwater.

conventional systems which are said to be costly and natural resource intensive. DWWT systems provide an integrated approach for diverse ecosystems with local applications at community level, individual level and institutional level.[35] They are site-specific and have high technical, social and economic feasibility (Box 8).

Based on the removal efficiency, land availability for treatment, availability of funds and power and skilled labor, some of the available technologies for wastewater treatment and reuse are summarized in Table 5.

4.2 Financial viability

Design and implementation of wastewater reuse projects requires a lot of planning and capital investment. A detailed analysis of the funding sources needs to be done before selecting any reuse scheme. In India, different funding sources may be available to an urban local body (ULB) for implementing such projects, broadly categorized as:

Table 5 A summary of some wastewater treatment technologies in India.[22,31,35,36]

Technology	Removal efficiency	F.coli removal	Energy requirement	Skilled labor requirement	Operation & maintenance cost	Capital cost	Reinvestment cost	Land requirement	Nutrient removal	Performance reliability
Activated Sludge Process (ASP)	Very high	Medium	Medium	Medium	High	Medium	Medium	Medium	Medium	High
Fluidized Aerated Bed Process (FAB)	Very high	Medium	Medium	Moderate	Medium	Medium	Medium	Low	Medium	High
Submerged Aerated Fixed Film Process (SAFF)	Very high	Medium	Medium	Moderate	Medium	Medium	Medium	Low	Medium	High
Trickling Filters	Very high	Medium	Low	Moderate	Low	Low	Low	High	Medium	Medium
Facultative Aerated Lagoons	Medium	Medium	Medium	Moderate	Low	Medium	Medium	Less	Medium	High
BIOFOR	Very high	Medium	Medium	Medium	Medium	Medium	Medium	Low	Medium	High
Sequential Batch Reactor (SBR)	Very high	High	Medium	Moderate	High	Medium	Medium	Low	Medium	Very high
Upflow Anaerobic Sludge Blanket (UASB)	High	Low	Medium	Medium	Medium	Medium	Medium	Medium	Low	Low
Cyclic ASP	Very high	Medium	Medium	Moderate	Medium	Medium	Medium	Low	Medium	High

Continued

Table 5 A summary of some wastewater treatment technologies in India.[22,31,35,36]—cont'd

Technology	Removal efficiency	F.coli removal	Energy requirement	Skilled labor requirement	Operation & maintenance cost	Capital cost	Reinvestment cost	Land requirement	Nutrient removal	Performance reliability
Waste Stabilization Ponds (WSP)	High	Medium	Very low	Very Less	Low	Very low	Low	High	High	High
Constructed Wetlands	High	Medium	Little/No	Less	Low	Low to Medium	Low	High	High	High
Moving Bed Bioreactor Process (MBBR)	Very high	Medium	Medium	Moderate	Medium	Medium	Medium	Low	Medium	High
Membrane Bioreactor (MBR)	Very high	Very high	Medium	High	High	High	High	Very low	Medium	Very high

- Government sources (such as the Government of India, State, and the Housing & Urban Development Corporation [HUDCO]).
- Domestic commercial sources (such as banks, insurance companies, and capital markets).
- International sources (such as multilateral financial institutions like the Asian Development Bank and the World Bank Group) and bilateral institutions such as the Japan International Cooperation Agency (JICA), GIZ, DFID.
- Public-Private Partnerships (PPPs) where a private entity takes on the responsibility of project implementation, including financing, and invests private equity or some sort of debt into the project.

Private sector engagement in designing and operating the reclaimed water reuse is an upcoming trend in India. Various multi and bilateral financial institutions are putting a lot of financial efforts for expansion of reclaimed water reuse in the country. It serves twofold purpose of technology transfer and capital investments. These recent initiatives have enhanced the financial feasibility of treated wastewater reuse projects in India. There are different PPP models for wastewater management and reuse in India as shown in Fig. 7. Out of the discussed financial models, build, operate and transfer (BOT) end-user and the design, build and operate (DBO) model have achieved varying levels of success in the Indian market.[31] Some of the successful PPP models for treated wastewater reuse in India are exhibited in Table 6.

4.3 Legal initiatives

The overall planning, implementation, and success of reuse projects depends highly on the effective policies, institutional framework and regulations. Environmental legislation plays an essential role in the adoption of reclaimed water reuse as part of water management. The long-term feasibility of integrated water management systems is reliant on sustainable guidelines and strict regulatory framework. Water sector in India is governed with policies of central and state Governments as well as local bodies governing rural and urban regions. Many initiatives have been taken by the authorities, at both central and state levels to promote the reuse of treated wastewater (Table 7). Lately, the water sector has seen several reforms in the water sector policies like, cost recovery, public-private partnership and water tariffs, etc., making way for integrated sustainable development including privatization of the sector.

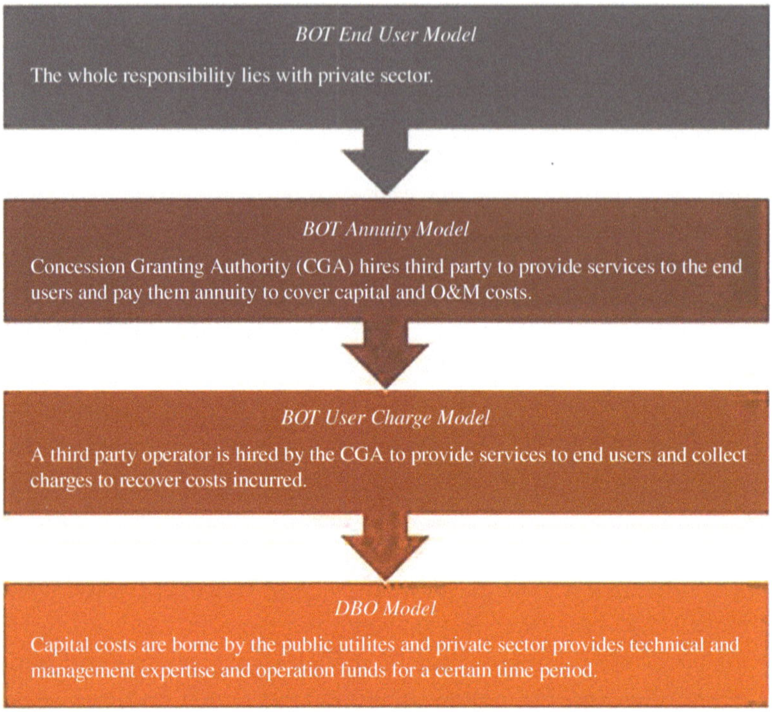

Fig. 7 Various financial models for wastewater management and reuse in India.

Table 6 An overview of PPP models for wastewater reuse in India.[31,37]

Name of the project	Type of model	Remarks
Vishakhapatnam Wastewater Recycling project	BOT User Charge	100% reuse of 60 MLD tertiary treated wastewater throughout the city, especially by industries
Surat—Bamroli Wastewater Recycling	BOT User Charge	Supply of approximately 40 MLD of tertiary treated water to industrial units in Pandesara Industrial Estate
Nagpur-MAHAGENCO (Maharashtra State Power Generation Company)	BOT End User	MAHAGENCO to receive entire 110 MLD plus 10% of treated wastewater from Pioli Nadi STP
Chennai Metro Water Supply and Sanitation Board (CMWSSB)	DBO Model	In 2016, CMWSSB collaborated on a PPP project to develop a reuse plant of 45 MLD capacity to supply non-potable water to industries

Table 7 Wastewater Reuse policies at National level.[13,22,23,32,38,39]

Government of India (Sector)	Policies	Objective
Ministry of Housing and Urban Affairs	National Urban Sanitation Policy, 2008	Endorses reuse of treated water, particularly, a minimum of 20% reuse of wastewater in every city
	National Mission on Sustainable Habitat, 2010	Promotes reuse by encouraging a dual-pipe system, separate metering, incentivization to customers, industrial reuse, reuse model development, and its propagation
	Service Level Benchmarking framework, 2010	Encourages reuse of at least 20% of treated wastewater
	Jawaharlal Nehru National Urban Renewal Mission, 2012	State governments and Urban Local Bodies of 63 cities, with central assistance, required to undertake mandatory and optional reforms to formulate bylaws for wastewater reuse Cities are provided assistance for 100% sewage treatment facilities to ensure better quality and increased availability of water for lower grade water use, e.g., Gujarat
	Guidance on recycle and reuse in the CPHEEO Sewerage and Sewage Treatment Systems, 2013	Incorporates a detailed discussion specifying the water quality guidelines of treated water for intended use, identifying best practices and case studies
	Swachh Bharat Mission, 2014	Encourages the reuse of treated wastewater by developing sewage collection and treatment networks
	Atal Mission for Rejuvenation and Urban Transformation, 2015	Providing basic services (e.g., water supply, sewerage, urban transport) to households and building amenities in cities to improve the quality of life
	Smart Cities Mission, 2015	Focuses on urban renewal and retrofitting of cities to make them sustainable and citizen-friendly
NITI Aayog	Water and waste management strategy in the 12th 5-year plan, 2013	Promotion of recycling and reuse of treated sewage in city planning and urban projects such as sanitation, landscaping, central air conditioning

Continued

Table 7 Wastewater Reuse policies at National level.[13,22,23,32,38,39]—cont'd

Government of India (Sector)	Policies	Objective
Ministry of Water Resources, River Development, and Ganga Rejuvenation	National Action Plan on Climate Change, 2008	Emphasizes on waste management and recycling. Set a goal of a 20% increase in water use efficiency through pricing and other measures
	National Water Mission, 2011	Increasing water use efficiency by 20% by promoting reuse
	National Water Policy, 2012	Recognizes the importance of reuse of water for meeting environmental goals and suggests preferential tariffs to incentivize reclaimed water
	Namami Gange Program, (2014–present)	Focus on Industrial effluent monitoring and sewage treatment to encourage reuse by the support of PPP models Agreement signed between NMCG and IOCL for reuse of 2 crore liters per day of treated sewage by IOCL's refinery in Mathura to reduce stress on the Yamuna MoU signed between NMCG and Ministries of Railways, agriculture, etc., for promoting reuse of treated wastewater Zero black liquor discharge achieved in paper & pulp industries with an estimated reduction of 45–50% in freshwater consumption compared to 2012 Effluent generation reduced from 400 to 200 L per tonne of sugarcane crushed The state of Bihar is drafting a policy of reuse of wastewater from STPs for irrigation Water reuse policy being drafted by Ganga basin states
	National Water Framework Law (Draft), 2016	Encourages implementation of economic models promoting sustainability of recycle-reduce-reuse using incentivization Encourages reuse of primary treated urban water effluents from kitchens and bathrooms in toilet flushing
Ministry of Environment, Forest and Climate change	National Environment Policy, 2006	Development of action plans to encourage projects, through public agencies and PPP for treatment, reuse, and recycle where applicable, for both industrial and municipal wastewater
Ministry of Power	Gazette Notification, 2016	Mandates the reuse of treated wastewater for all thermal power plants located within a radius of 50 km from any city, e.g., Nagpur

4.4 Market viability

The successful reuse of treated wastewater can be accomplished only if the treated water has a market potential. Market viability basically implies ascertained end-users for the reclaimed water and economic sustainability of the project. It depends on several factors as presented in Fig. 8.

The treated wastewater has a very high applicability in Indian market as reported with the following instances.

- Currently, many cities transport their water from distant sources leading to high conveyance costs, e.g., Chennai acquires water from Lake Veeranam which is 235 km from the city; Bangalore gets the water from 95 km away Cauvery River and needs to be pumped at an elevation of 1000 m; Hyderabad sources its water from the Krishna River situated at a distance of 130 km from the city which requires expensive multi-stage pumping, etc.[13] Use of treated wastewater will help in reducing the water costs.
- Central pollution control board (CPCB) in India has implemented zero liquid discharge (ZLD) on all water polluting industries in India, which has made water recycling and reuse mandatory in industries.
- Research shows that using treated wastewater for agriculture has many monetary benefits.
- Reclaimed water has become a source of revenue for many public utilities. Treated water is sold to industries which covers certain costs, making it an attractive water supply option. For example, in Chennai,

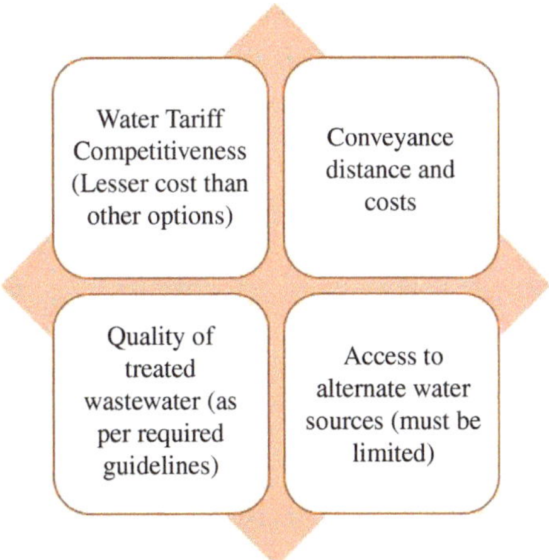

Fig. 8 Factors impacting market viability of reclaimed water reuse projects.

treated wastewater is sold to industries at INR 8–11/m^3; Chandigarh municipality charges INR 500/acre for supplying treated wastewater for agricultural irrigation and charges INR 50/500 $yards^2$/month for irrigation of green spaces.[33]

5. Conclusions

Population growth and economic development has led to an ever-increasing water demand in India. Limited potential of existing water resources has further strained the issue. The past efforts to close the widening water supply-demand gap in the country have been too little and too late. Alternative water augmentation strategies, particularly, the use of treated wastewater has proven to be a promising solution to deal with the water crisis. Studies show that the reuse of the treated wastewater is the most viable option as it has many environmental and economic benefits with minimal resource utilization.

Treated wastewater reuse is still an evolving concept in India. Many governing policies and regulatory frameworks exist at the central government level, but the adjustment and implementation at state and local levels still need a lot of push, planning and awareness. Due to certain modifications in environmental legislation, the industrial reuse market capacity has grown significantly in the past few years. Even though, there are a lot of opportunities that encourage the reuse of treated wastewater in the country, they are often squandered.

To achieve sustainable reuse of treated wastewater, there is a need to develop and explore funding and revenue options through different business models, technical infrastructure and potential users of reclaimed water. The design and operation and maintenance considerations for the reclaimed water reuse projects and tariff models for sale of treated water for different applications needs to be developed. There is a strong need to form a national reclaimed water reuse policy with targets, regulatory and financial measures. Further, the existing water laws need reforms and expansion to address the diminishing per capita availability, quality and access of freshwater.

Additionally, to make reclaimed water reuse a success in India, social factors like education levels and public perceptions on the reclaimed water and their awareness toward the environment must be considered. Their willingness and ability to pay and protect the environment plays an important role in designing treated wastewater reuse systems. The political will and institutional support is also essential. Considering the Indian scenario, the following conclusions may be drawn.

(1) There is a large scope for advancement of treated wastewater reuse in the country along with agricultural and industrial reuse.

(2) To gain mainstream attention for reclaimed water reuse, it needs to be included in the existing and upcoming water supply and wastewater management programs through sustainable policy and regulatory framework and effective institutions, especially in water sensitive regions. A comprehensive policy at national and state level is required.

(3) There is a need to promote ongoing treated wastewater reuse efforts and encourage future endeavors, especially in the cities developing sanitation plans, among the policymakers, donors, civil societies and other stakeholders.

(4) Policy reversal from the perspective that water is a public resource to the introduction of economic value of water entitled for trading. Transforming the role of the government by transferring part of existing governmental prerogatives to users and private sectors would assist in enhancing the financial feasibility of reuse projects.

(5) Shifting the focus from supply side management to demand side management would lead to improvement in water planning and management.

(6) Formulation of a holistic approach would help the municipalities to meet the future needs of an ever-increasing population, industrialization and economic growth.

References

1. United Nations World Water Development. *Nature based solutions for water.* 2018, The United Nations World water development Report.
2. *Central water commission. Water—its conservation, management and governance.* Government of India; 2017.
3. *Central water commission. Annual report 2015–16.* Government of India; 2016.
4. *Central pollution control board. Annual report 2014–15.* Government of India; 2015.
5. NITI Aayog. *Composite water management index—a tool for water management.* Government of India; 2018.
6. McNabb DE. *Global pathways to water sustainability.* Springer; 2019.
7. Arceivala SJ. *Design construction and operation of waste stabilisation ponds in India.* 1969.
8. Arceivala SJ, Asolekar SR. *Wastewater treatment for pollution control and reuse.* Tata McGraw-Hill Education; 2015.
9. Ravikumar J, Raman R, Jain S. *Wastewater recycle and reuse: an economical and sustainable option [Internet].* [cited 2019 Jan 31]. Available from http://mohua.gov.in/upload/uploadfiles/files/5_WSP Presentation_WWR_Mar 201409.pdf; 2014.
10. Dutta S, Bhaskar S. *Surface and groundwater pollution are pushing India towards a water crisis.* Available from: https://swachhindia.ndtv.com/world-water-day-2018-surface-and-groundwater-pollution-are-pushing-india-towards-an-irreversible-water-crisis-17882/; 2018.
11. *Central ground water board. Annual report 2016–17 [Internet].* Government of India, 2017, Available from: http://cgwb.gov.in/Annual-Reports/Annual Report 2015-16.pdf.

12. *India: second biennial update report to the United Nations framework convention on climate change.* Ministry of Environment, Forest and Climate Change, Government of India; 2018.
13. *Recycling and reuse of treated wastewater in urban India: a proposed advisory and guidance document.* CGIAR Research Program on Water, Land and Ecosystems. International Water Management Institute (IWMI); 2016.
14. *The reuse opportunity. The international water association. [Internet].* Available from http://www.atmos-chem-phys-discuss.net/acp-2016-176/; 2018.
15. George BA, Malano HM, Khan AR, Gaur A, Davidson B. Urban water supply strategies for Hyderabad, India future scenarios. *Environ Model Assess* 2009;**14**:691–704.
16. Barringer J. Urban domestic and commercial water reuse in Pune and its influence on the present water crisis. *Water Qual Expo Health* 2014;**6**:35–8.
17. Ghosh R, Kansal A, Venkatesh G. Urban water security assessment using an integrated metabolism approach-case study of the National Capital Territory of Delhi in India. *Resources* 2019;**8**(2):1–15.
18. *Review of wastewater reuse projects worldwide collation of selected international case studies and experiences. GRB EMP: Ganga River basin environment management plan.* Indian Institutes of Technology; 2011.
19. Buechler S, Mekala GD. *Household food security and wastewater-dependent livelihood activities along the Musi River in Andhra Pradesh, India [Internet].* Available from: http://www.who.int/water_sanitation_health/wastewater/gwwufoodsecurity.pdf; 2005.
20. Buechler S, Mekala GD, Raschid-Sally L. Livelihoods and wastewater irrigated agriculture along the Musi River in Hyderabad City, Andhra Pradesh, India. *Urban Agric Mag* 2002;**8**:14–7.
21. Mekala GD, Davidson B, Samad M, Boland A-M. *Wastewater reuse and recycling systems : a perspective into India and Australia.* International Water Management Institute; 2008.
22. Shah M. *Urban water systems in India : a way forward.* Indian Council for Research on International Economic Relations; 2016.
23. Alley KD. Parameters of successful wastewater reuse in urban India. *Indian Polit Policy* 2018;**1**(2):91–122.
24. Devi MG, Samad M. Wastewater treatment and reuse: an institutional analysis for Hyderabad, India. *Water* 2008;513–23.
25. *Notification for not using treated wastewater for agriculture dated April 15, 2015.* India: Uttar Pradesh Jal Nigam; 2015; 63–75.
26. (CPHEEO) Central Public Health & Environmental Engineering Organisation. *Recycling and reuse of sewage. Manual on Sewerage and sewage treatment systems.* CPHEEO; 2013.
27. Winrock International India. *Urban wastewater: livelihoods, health and environmental impacts in India.* Winrock International India; 2007.
28. Mekala GD. Multi-stakeholder processes for safe use of wastewater and sustainable urban, peri-urban agriculture—a case study of Hyderabad, India. In: *Powerpoint presentation at the UN Habitat third World Urban Forum.* 2006; 19–23.
29. Delhi Jal Board. *Action plan for utilization of treated effluent.* Government of Delhi; 2019.
30. *Recycling and reuse of sewage [Internet].* cited 2019 Jan 4. Available from: http://cpheeo.gov.in/upload/uploadfiles/files/engineering_chapter7.pdf; 2019.
31. *Urban wastewater public-private partnerships.* FICCI Water Mission and the 2030 Water Resources Group (2030 WRG); 2016.
32. *National workshop on wastewater recycle and reuse [Internet].* Available from http://mohua.gov.in/upload/uploadfiles/files/NATIONALWORKSHOPONWASTEWATERRECYCLEANDREUSE_BackgroundNote02.pdf; 2014.
33. Lautze J, Stander E, Drechsel P, da Silva AK, Keraita B. *Global experiences in water reuse.* CGIAR Research Program on Water, Land and Ecosystems (WLE); 2014.

34. Chauhan S, Jensen O, Sengaiah G, Screenivas N. *Closing the water loop: reuse of treated wastewater in urban India*. (September). Available from: https://www.pwc.in/assets/pdfs/publications/2016/pwc-closing-the-water-loop-reuse-of-treated-wastewater-in-urban-india.pdf; 2016.
35. Matto M, Dwivedi D, Sharda C, Tallu A. *Decentralised wastewater treatment and reuse—case studies of implementation on different scale—community, institutional and individual building [Internet]*. Available from http://www.databank.com.lb/docs/Water-SustainableManagementofaScarceResource-AFED-2010.pdf#page=99; 2014.
36. *Draft guidance note for promoting municipal wastewater reuse and reclamation in India*. Japan: NJS Engineers India Pvt. Ltd. (A wholly owned subsidiary of NJS Consultants Co., Ltd; 2015.
37. *State policy for wastewater reuse for Jammu & Kashmir [Internet]*. Available from https://www.indiawaterportal.org/sites/indiawaterportal.org/files/wastewater_reuse_policy_for_jk_state-_10-04-2017_final.pdf; 2017.
38. *Gazette notification for using treated wastewater in thermal power plants dated January 28, 2016*. Ministry of Power, Government of India; 2016; 38.
39. Freedman J, Meeker JTM, Vallero M. *Addressing water scarcity through recycling and reuse : a menu for policymakers*. WateReuse Association and Research foundation; 2014.

CHAPTER THREE

Water reuse practices, solutions and trends at international

Sylvain Donnaz*
Suez International, Rueil-Malmaison, France
*Corresponding author: e-mail address: sylvain.donnaz@suez.com

Contents

1. Water Reuse Practices Worldwide: "Not if, but how" 66
2. On the rise of water reuse worldwide 68
3. A brief review of water reuse technologies and solutions 72
4. The case of aquifer recharge in the value chain 77
   4.1 How to integrate MAR in water resource and treatment projects? 80
   4.2 Treatment solutions to be considered for Aquifer Pre-treatment and Post-treatment applications 80
5. Pioneering References and Flagship Water Reuse Projects around the World 84
6. Horizon 2040 of water reuse 95
   6.1 Indirect potable reuse coupling wastewater reuse with managed aquifer recharge 95
   6.2 IPR and the rise of Direct Potable Reuse 96
   6.3 Agriculture 96
   6.4 Micropollutants Removal to Protect environmental bodies and fauna 98
   6.5 Discussion about Water Reuse in Europe 99
Acknowledgments 100
References 100

Abstract

Water Reuse is a reality at international. Several practices have emerged around the world as results of different needs, perspectives and policies. Accelerating development occurred in the past 15 years, especially in the application field of potable reuse. Technologies and treatment flowsheets have been studied and validated, demonstrating advanced water quality, safety and suitability for potable reuse applications. Indirect Potable Reuse has known recent development while Managed Aquifer Recharge coupled with Water Reuse plays a key role in water management and infrastructure projects (brownfield or greenfield) when considered at conceptual and feasibility study stages. Most recent infrastructure designed, built and operated by private process integrators demonstrate the increasing development of treated wastewater reuse coupled with direct artificial recharge, in terms of technologies but also in terms of regulation. Aquifers are more than ever the nexus of Indirect Potable Reuse to increase global water

Advances in Chemical Pollution, Environmental Management and Protection, Volume 6
ISSN 2468-9289
https://doi.org/10.1016/bs.apmp.2020.07.012

resource, recharge local water cycle, centralize water monitoring and provide sustainable solution in many regions. Pioneering References and Flagship Water Reuse Projects around the World demonstrate how wastewater reuse infrastructure investments are safe and constituting a sustainable response to water resource planning in the context of climate change. Protection of environmental bodies and fauna, fit-for-purpose agriculture reuse and indirect potable reuse coupling wastewater reuse with direct aquifer recharge are key water reuse schemes for the future.

Keywords: Wastewater reuse, Indirect potable reuse, Managed aquifer recharge, Groundwater recharge, Direct aquifer recharge

1. Water Reuse Practices Worldwide: "Not if, but how"

A compelling criterion for water planning is the anticipation of water resource management at the boundaries of regional and local water cycles. As snowpack decrease all over the world in the mountain regions, storages of autoregulated natural water volumes tend to disappear. At lower altitudes downstream and in plains, groundwater remains the last line of defense against droughts. During scarcity, people rely more heavily on groundwater[1]; therefore, groundwater depletion increases worldwide.[2] There is a decrease of natural recharge combined with the rise of withdrawals. Those effects find its roots in the human impacts and Climate Change.

One of the consequences of climatic change is annual precipitation reduction accompanied by intensity increase during rainy season in many regions of the world where Managed Aquifer Recharge techniques and water recycling will be a solution (Fig. 1).

In this context, the adoption of Water Reuse for long term water planning is key, when applied to the boundaries of regional and local water cycles.

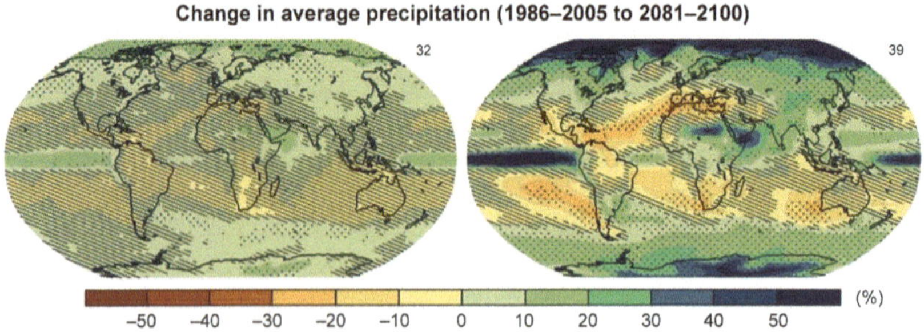

Fig. 1 Change in average precipitation (1986–2005 to 2081–2100).[3]

Water Reuse shall be seen from two different perspectives: from the "resource" and from the "final usage" perspectives. In the professional community, Water Reuse designates all practices that involves reusing water that is not coming from the conventional drinking water supply and that is used for different purposes. From the resource perspective, it includes storage and reclaimed water from rainfalls, surface waters and reused water from municipal and industrial treated wastewaters. From the "final use" perspective, it includes:

- Non-potable use
  - Urban usages such as road cleaning, recreational and environmental uses such as artificial lakes, ornamental fountains, wet lands...
  - Landscape and green areas' irrigation
  - Agriculture irrigation
  - Industrial process waters for various uses such as cooling towers, pressure boilers...
- Potable reuse, direct or indirect, involving the production of drinking water for storage (surface or groundwater) and/or distribution for human consumption

Reusing treated wastewater designated by "Wastewater Reuse" does consist in the better integration of "human activities" in the natural water cycles (global, regional and local). With this concept, treated wastewater appears as part of the available water resources, allowing to preserve and decrease pressure on natural water resource bodies. Instead of letting treated wastewater being discharged, transferred and reused *de facto* far away from its point of use, this water can be re-processed again through tailored treatment technics and redistributed and reused locally. All in all, treated "Wastewater Reuse" shortens the natural water cycle in order to better preserve the environmental water resources put at stress by human demographic and activities pressures. "Wastewater Reuse" improves water management in terms of quantity and quality. It kills two birds with one stone, by complementing the lack of resources and contributing to replenish the natural stocks, it also improves the quality of treated water by means of additional advanced and polishing treatment stages required for its usages.

There are two types of main usages, which implies different technological solutions:

- "Easy Reuse," which encapsulates irrigation of green areas, mainstream agriculture irrigation, recreational and environmental enhancement usages (artificial lakes, leisure parks, wet natural zones...). Urban usages such as road and municipal vehicles cleanings are part of it

Table 1 Wastewater Reuse Practices Worldwide 1990–2014.

Reuse practices worldwide		Easy/advanced	Tendency (2020onward+)
Agriculture irrigation	32%	75%	++
Landscape irrigation	20%		+
Recreational and environmental usages	15%		0
Urban usages	8%		0
Industry	20%	25%	+++
IPR	4.5%		+++
DPR	0.5%		++

- "Advanced Reuse," which includes process water for industries (cooling towers, high-pressure and low-pressure water boilers, air conditioning…), groundwater storage and resource augmentation for drinking water supply (IPR) and potable water production and supply through Direct Potable Reuse. This also includes fit-for-purpose agriculture irrigation where specific water quality must be produced for sensitive crops and plants, e.g., for the control of salinity (TDS) or dissolved solids

Considering the above categories, Wastewater Reuse practices worldwide (1990–2014) are made of 75% with Easy Reuse (Municipal usages); 25% Advanced Reuse (Industrial and Municipal), as summarized in Table 1.

2. On the rise of water reuse worldwide

In the past 30 years, Water Reuse has been encapsulated under several names, from the misfit "toilet-to-tap" to the virtuous "use of recycled water," "treated wastewater" or "wastewater reuse." Several practices have emerged around the world as results of different needs, perspectives, drivers and policies.

Looking back to the last decades and considering current and future path forward, four major types of Water Reuse can be identified and proposed today:

- Basic Reuse or "*De Facto* Reuse";
- Tertiary Reuse;
- Advanced Tertiary Reuse and
- Multibarrier Reuse

Basic Reuse or "De Facto Reuse" is the current situation about surface waters where interactions between upstream and downstream exist. This occurs when Wastewater Treatment Plants (WWTP) discharge treated wastewater upstream into rivers, where several kilometers downstream, Drinking Water Plants pump and use river water as a source for potable water production and supply. Supranational regulations in the early 90s (like the Water Directive set by the European Union) have greatly contributed to improve the quality of surface water masses by reinforcing the discharge limits in terms of carbon, nitrogen (Total Nitrogen 10 mg/L) and phosphorus (Phosphorus 0.2–2 mg/L). Large investments have allowed the construction and upgrade of WWTP across Europe (1990–2020) that ensured the improvement and protection of water receiving bodies, especially surface waters. Those investments have largely been used to build and extend secondary biological treatment capacities.

Tertiary Reuse has developed in specific areas where WWTP discharge interacts with other human activities (bathing areas, green areas like golf or public parks, agriculture irrigation). In those applications, the need to further improve the Total Suspended Solids (TSS) removal for efficient and sustainable water irrigation spraying comes with the need to ensure Pathogens removal to prevent human bodies from any biological contamination and disease. In Tertiary Reuse applications, investments concern the design, construction and operation of tertiary filtration (sand filters, ultrafiltration membranes) coupled with final disinfection units (chlorination, chlorine contact tanks, ultra-violet). This type of reuse has developed worldwide in specific cases for easy irrigation applications such as mainstream agriculture and green areas (golf courses) in Southern Europe (Spain, Italy), West coast in Northern America (e.g., the early Title 22 California regulation), Central and Latin America (Mexico, Chile...).

Advanced Tertiary Reuse deals with the implementation of advanced disinfection coupled with the need to implement either very low Chemical Oxygen Demand (COD) levels, Total Organic Carbon (TOC) removal and micropollutants removal, which includes Compounds of Emerging Concern (CEC) like pharmaceuticals, endocrine disruptors, pesticides... This level of treatment generally includes combination of different solutions such as tertiary ozonation, activated carbon contactors (PAC: Powder Activated Carbon or GAC: Granular Activated Carbon), Biologically Active Filtration (BAF) coupled with ozone... In China, the new Grade IV regulation does impose to achieve very low Biological Oxygen Demand (BOD) and COD levels ($BOD_5 < 6\,mgO_2/L$; $COD < 30\,mgO_2/L$) for surface water quality. Grade III of the Chinese regulation goes even beyond with BOD < 4 mgO_2/L and COD < 20 mgO_2/L. This is also the case of Switzerland where pioneering micropollutants regulation has been implemented from the kick-off year of 2016 to the horizon of 2040. Micropollutants removal implies the construction of new advanced tertiary treatment stage in more than 120 WWTPs out of 650. In Europe, more and more countries such as Denmark, Sweden, Germany and France are preparing new micropollutants regulation, which will foster the implementation of advanced tertiary treatment for micropollutants during the coming decades (2020–2040).

Multibarrier Reuse is the utmost water reuse category, which main objective is to produce Potable Water supply, be it Indirect Potable Reuse (IPR) or Direct Potable Reuse (DPR). The objective is to ensure utmost levels of Pathogens removal, TOC removal and when necessary Total Dissolved Solids (TDS) removal. Such applications imply the use of membrane technologies (mainly Ultrafiltration: UF; Reverse Osmosis: RO), PAC/GAC filters, contactor or reactors, ozone coupled with BAF, Reverse ElectoDialysis (EDR), UV coupled with Advanced Oxidation Processes (AOP)... all advanced treatment technologies combined to ensure the highest level of removal and redundancy. Those applications are led by the US-Californian and Australian regulations. For example, the triple barrier concept for Pathogens removal, Log RemoVal (LRV) credits for virus, bacteria and protozoa (including Cryptosporidium, Giardia...) applied in USA for IPR and DPR.

Table 2 summarizes the different types of Wastewater Reuse proposed in terms of treatment objectives, drivers, regulation and in which lead countries it is implemented.

Table 3 provides Key Performance Indicators that may be used to assess the treatment performance of different wastewater reuse treatment flowsheets. This table is a simple guideline on how to evaluate and compare different treatment lines for decision making prior to project implementation.

Table 4 provides magnitudes of water prices depending on the type of water.

Adoption of Water Reuse globally has occurred. Since 2015, Wastewater Reuse has become the real alternative to desalination when it comes to solving water scarcity issues. In the last 5 years, applications of water reuse for Potable use (Indirect and Direct) has significantly increased.

Table 5 gathers different Total Organic Carbon levels that have been chosen as guarantee limits to regulate potable water or potable reuse projects.

Table 2 Water Reuse: a matter of perspectives and drivers worldwide.

	Major treatment objectives	Drivers	Policy/Regulatory approach	Territories
Basic Reuse or *De Facto*	Nitrogen, phosphorus	Water masses and surface water improvement	Supranational (European Directive), National, Federal or State	EU, USA
Tertiary	TSS, pathogens	Bathing areas, irrigation in agriculture or landscaping	Supranational (European Directive), National, Federal or State	EU, USA
Advanced tertiary	Pathogens, TOC	Water scarcity, protection of environment and fauna	National	China, Switzerland, USA, South EU
Multibarrier	Pathogens, TOC, TDS	Water scarcity, potable reuse indirect or direct, fit-for-purpose industrial and agricultural uses	Regional, Local	USA, Australia, Singapore, Taiwan

Table 3 Key Performance Indicators (KPI) for Water Reuse treatment lines.

Performance indicators		Scoring of treatment line Alternative 1	Scoring of treatment line Alternative 2
TSS			
TOC removal			
Pathogens Log RemoVal (LRV)	Bacteria		
	Virus		
	Protozoa, Crypto, Giardia		
TDS removal			
Micropollutants or CECs removal			

Table 6 describes current practices of wastewater reuse per country in terms of volumes of recycled wastewater and technological level used for treatment.

Table 7 provides wastewater recycled volume breakdown by state (USA) and countries (Europe):

Table 4 General water production (selling prices[a]).

Origin of water	USD/m^3
WWTP	0.15–0.30
Reclaimed water (Easy Reuse)	0.25–0.50
Sea water desalination	0.50–1.00
Drinking water (plant + network)	1.00–1.50
Bottled water	800–1500

[a]Prices function of treatment technologies, capacities and local market conditions (resource, usage, demand).

Table 5 Different approaches on Total Organic Carbon (TOC) limits worldwide.

Treated water std	Parameter	Value		Application
Disinfection	TOC in mg/L	<4.0	Disinfection	
USA, Virginia		<3.8	Reuse-IPR	Aquifer recharge
USA, Florida		<3.0	Reuse-IPR	Direct injection into aquifer
USA, EPA Guidelines		<2.0	Reuse-IPR	Direct injection into aquifer
USA, California		<0.5	Reuse-IPR	Direct injection into aquifer
WHO		<2.0	Drinking Water	
EU Directive		<2.0	Drinking Water	
Australia, Perth		<0.05	Reuse-IPR	Direct injection into aquifer
Mexico		<1.0	DW and Reuse-IPR	Direct injection into aquifer

3. A brief review of water reuse technologies and solutions

Over the 1990–2010 period, the emergence of membranes has largely contributed to the development of Water Reuse applications around the world, from 2015 to 2020, alternatives treatment trains started to be seriously considered, both at academic and business applications levels.

Table 6 Recycled wastewater volumes and technological levels.[4]

	% Recycled wastewater (vol/vol)	Technological level
Mexico	78% including 40% only *with treatment*	0
Israel	70% (100% in Tel Aviv)	++
Kuwait	35.2%	+
Tunisia 2020 objective	18%	0
Europe (Top 3)	2.4%	0
Spain	15%	++
Italy	5–15%	0
Greece	5%	0
USA	<1%	+++

Table 7 Wastewater recycled volumes in USA by state and countries in Europe.

USA in MGD/year	
Irrigation for agriculture	**Europe in Mm³/year**
1. California (270)	1. Spain (347)
2. Florida (256)	2. Italy (233)
3. Arizona (23)	3. Germany (42)
4. Texas (19)	4. Greece (23)
5. Nevada (13)	5. The Netherlands (7.6)
6. Colorado (3)	6. Benelux (2.9)
7. Caroline du Nord (1)	7. Poland (2.6)
8. Wyoming (0.89)	8. Portugal (0.3)
9. Utah (0.81)	
10. Idaho (0.27)	
11. Washington (0.02)	

"All membrane" solutions such as Ultrafiltration (UF) followed by Reverse Osmosis (RO) and Ultraviolet (UV) have widely been used in potable water reuse applications, allowing projects to be development thanks to highest sanitary certification levels in terms of disinfection, removal of trace organic compounds and micropollutants. After 20 years, if advantages brought by "all membrane" solution are still largely acknowledged in terms of water quality and process reliability, the fact that membranes divert and produce byproduct such as brine (salt and micropollutants rejection) start to generate operational issues. Pressure increases on operations which face operational management issues with brine permitting, disposal and transportation. Since 2015, the professional community started to envision different alternatives and to compare "all membrane" solution to "Ozone-Biologically Active Filtration" in terms of treatment performance, capital and operational expenditures. Both solutions present advantages and drawbacks that are "project specific" and that shall be assessed case by case (Table 8).

Studies piloting comparison for in land potable reuse between UF-RO-UV and O_3-BAF-UV have shown that both treatment flowsheets achieve similar performances (Jacksonville, Florida, USA). According to CDMSmith,[5] "*pioneering study enhances the potable reuse industry's grasp of two leading technologies by testing them side-by-side at two different WRFs. The Ozone-BAF-AOP treatment process, while widely used in the drinking water industry, is just beginning to warrant consideration in the potable reuse arena. The primary advantage of the Ozone-BAF-AOP treatment system compared with UF-LPRO-AOP is that Ozone-BAF-AOP does not produce concentrate. The cost and*

Table 8 Discussion: membranes or O_3-BAF?

"All membrane"	**Ozone-biologically active filtration**
UF-RO-UV	**O_3-BAF**
Pathogens removal (=) Micropollutant removal (=)	Pathogens removal (=) Micropollutant removal (=) when combined with activated carbon (AC)
Early market acceptance	Increase market acceptance
Diversion of CEC into brine	Removal of CEC into atmosphere (μgCO_2)
Need to manage brine disposal and CEC destruction	No brine, even more sustainable when coupled with environmental buffer (aquifer polishing)
Sea shore needed for ocean discharge permitting	Feasible, even Inland
Over quality (TOC < 1 ppm; not even necessary, water not even drinkable at this level)	Flexible function of operational water quality Better OPEX efficiency (upon needs: TOC fluctuations)
Discussion for IPR scheme: fine when coupled with Aquifer residency time <2 months	Discussion for IPR scheme: fine when coupled with Aquifer residency time >2 months

environmental concerns associated with concentrate management can impose limitations for some utilities. The Ozone-BAF-AOP treatment process, while widely used in the drinking water industry, is just beginning to warrant consideration in the potable reuse arena. The primary advantage of the Ozone-BAF-AOP treatment system compared with UF-LPRO-AOP is that Ozone-BAF-AOP does not produce concentrate. The cost and environmental concerns associated with concentrate management can impose limitations for some utilities (…) UF-LPRO-AOP met and exceeded all water quality goals, including primary and secondary drinking water standards and water quality goals for currently unregulated parameters. Ozone-BAF-AOP met primary and secondary drinking water standards and water quality goals for unregulated parameters at Southwest WRF. The Ozone-BAF-AOP system was challenged with meeting several secondary drinking water standards at Buckman WRF, including manganese, total dissolved solids and chloride at Buckman WRF."

Best Available Technologies (BAT) for Advanced Tertiary and Multibarrier Reuse are summarized in Fig. 2.

While Biological Nutrient Removal (BNR) ensures carbon, nitrogen and phosphorus removal, the need to achieve further and lower Total Phosphorus (TP) and Total Suspended Solids (TSS) levels is key for water reuse. This level of performance is reached with Tertiary treatment (Mechanical microfiltration, UF, sand filtration) made of BNR coupled with tertiary or of Membrane Biological Reactor (MBR).

Once TP and TSS levels are satisfying, water treatment engineers will deal with optimizing disinfection level and the removal of dissolved organic contents. Total Organic Carbon (TOC) concentration will often be used as

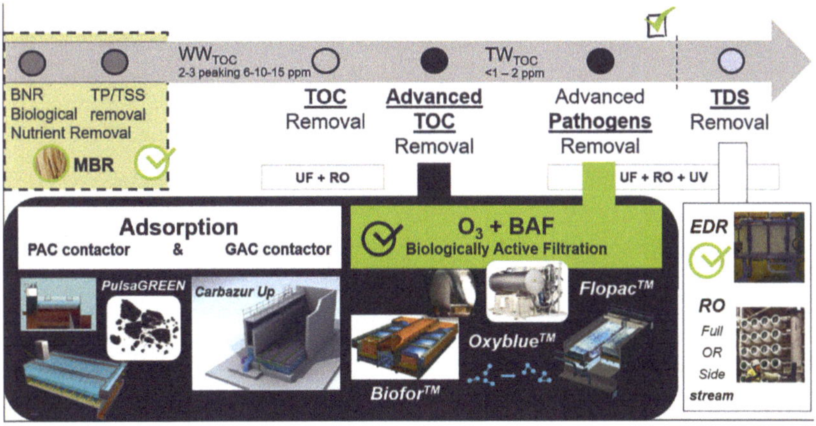

Fig. 2 Technologies for Multibarrier Reuse applications, Potable Water compliance and MIcropollutant/CEC removal.

non-specific surrogate of the presence of micropollutants (the lower TOC, the better quality). Advanced TOC removal can be achieved with membranes only by using RO (TOC < 0.25 mg/L after RO[6]). The contribution of TOC removal from UF is minimal.[7] When considering alternative solution using non-membrane process, advanced TOC removal can be achieved by Ozone-BAF and Adsorption processes.

The World Health Organization (WHO) provides clear guidelines for Log Removal (LRV) performances required for potabilization. It specifies pathogens occurrence and removal requirements for a list of selected indicators and pathogens (bacteria, virus, protozoa) in different water qualities, including feces, untreated wastewater and raw water for Drinking Water applications.

When it comes to pathogens removal performances in wastewater reuse applications, US California and Western Australia are the leading-edge references (Table 9).

Controlling the salinity level in wastewater reuse is key for both potabilization and agriculture. Frequency and levels variation of salinity in

Table 9 Comparison of pathogens and contaminants reduction in California and Western Australia IPR approaches.[6]

Requirement	California	Western Australia
UF/MF	0 to 1-log virus 4-log Giardia 4-log Cryptosporidium Turbidity < 0.2 NTU	3-log virus 3-log Cryptosporidium 3-log Campylobacter Turbidity < 0.2 NTU
RO	1 to 2-log virus 1 to 2-log Giardia 1 to 2-log Cryptosporidium TOC < 0.25 mg/L	3-log virus 3-log Cryptosporidium 3-log Campylobacter
UV	6-log virus 6-log Giardia 6-log Cryptosporidium	4-log virus 4-log Cryptosporidium 4-log Campylobacter
AOP	0.5-log 1,4-dioxane	Not required
Total pathogen requirement	12-log virus 10-log Giardia 10-log Cryptosporidium	9.5-log virus 8-log Cryptosporidium 8.1-log Campylobacter

wastewaters can cover a wide range, very dependent on local geology, drinking water source and sewage network infiltration, especially in coastal areas where sea intrusion occurs. Therefore, the need to implement Total Dissolved Solids (TDS) removal will be function of the specific quality of the treated wastewater and on its expected final use(s).

In the case of Potable Reuse, the World Health Organization (WHO) specifies for drinking water production that Total Dissolved Solids (TDS) level shall be lower than 600 mg/L to ensure good palatability of water. With TDS levels "*greater than 1,000 mg/L, drinking water becomes significantly and increasingly unpalatable.*"[8]

In the case of agriculture reuse for irrigation, the level of salinity will be highly dependent on the type of crops, plants and soils. High TDS concentrations can burn leaves, limit plants growth, impact crops production efficiencies and lead, in specific conditions, to sterilize soils in long term. Therefore, in the case of agriculture reuse, the compelling need to specify TDS level in the final treated water will directly impact the growth rate of the cultures, the economy and the sustainability of the system. There is a crucial need for more project development promoting fit-for-purpose agriculture irrigation qualities with specific prescription of salinity concentration levels to be achieved. This alone can ensure that suitable water treatment solutions will be implemented. Too often, salinity levels are not considered in wastewater reuse irrigation projects.

4. The case of aquifer recharge in the value chain[9]

At international, municipal water utilities in collaboration with private companies in the water sector have historically contributed to the development of artificial recharge. Managed Aquifer Recharge (MAR) has been implemented as early as 1965–70 in France and Spain. Its development accentuated in 1990 and accelerated after 2000 in Australia and United States in the context of Indirect Potable Reuse (IPR) applications.

Managed Aquifer Recharge (MAR) is at the center of the value chain for water resource management and can be coupled with different sources of water with key water treatment infrastructure. Confined or unconfined aquifers can be recharged directly or indirectly through various types of water: surface water, desalinated water and treated wastewater. MAR can be implemented for different purposes to: replenish overexploited aquifers, protect aquifer against saline intrusion, implement water storage, recover water quality through aquifer's complimentary soil treatment.

Coupling treated wastewater reuse (or Water Reuse) with artificial aquifer recharge is an efficient and competitive solution compared to the investment in water storage in traditional civil structures and other options that require treatment and transportation over long distances. Therefore, MAR coupled with Water Reuse plays a key role in water management and infrastructure projects (brownfield or greenfield) when considered at conceptual and feasibility study stages.

Wastewater Reuse for Managed Aquifer Recharge is key for the management of local water resource and often has a strong contribution within the local water cycle. It replenishes water reserves, acts as hydraulic fence/barrier against pollution risk, decreases water losses by evaporation. It is always the last line of defense against drought. If planned wisely, it is a solution for anticipating crisis situations and water scarcity with significant beneficial collateral costs benefits.

When MAR is integrated in water planning, the water treatment requirements depends on supranational, national, regional or local regulatory frameworks. Solutions will be function of environmental and sanitary safety levels, from hydrogeological characteristics of the aquifer up to the need to guarantee Potable Water standards (Fig. 3).

In France, private companies have designed, built and operated direct/indirect aquifer recharge applications. In Lyon, indirect recharge has been implemented to protect the alluvial groundwater from the risk of pollution intrusion of the Rhone river. In Croissy-sur-Seine (western Paris), direct recharge has been implemented after long period of over-exploitation to secure the water resource by ensuring groundwater refill. Seine river has been diverted and used, after water treatment, for direct injection. Groundwater has contributed up to 50% of the drinking water supply for the production and distribution of 500,000 inhabitants.

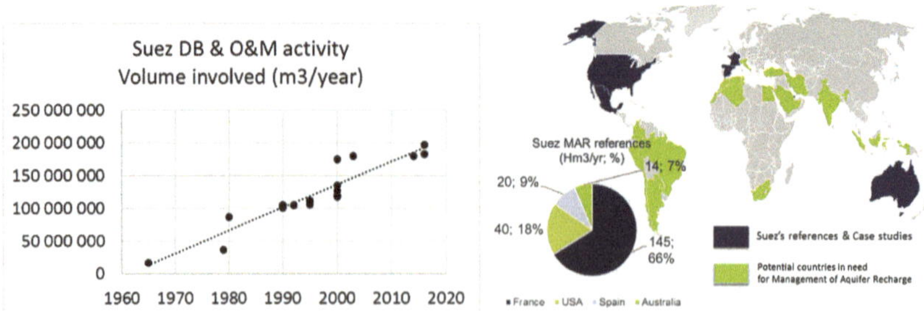

Fig. 3 MAR projects implemented by Suez worldwide (surface water and wastewater reuse).

In Spain, during the 1970s in the Delta del Llobregat of Barcelona, the MAR system included the water treatment before injection, reaching quality standards for drinking water. This application was the result of a wide and deep study to prospect new water resources for the Barcelona metropolitan region supply while long term planning of the hydric resources in the surrounding area. Thanks to these works the geometric and hydrodynamic knowledge of the aquifer was improved and the first flow model was developed. MAR was developed at a time when no alternative resources existed, as for example the seawater desalination and reclaimed water reuse, contributing in this way to secure the water supply in the growing urban and industrial area. Much later, in years 2000, the MAR project of the Señorío de Marbella aquifer allowed satisfying the demand during summer months. For short periods of time, the aquifers in this region was suffering an intensive operation and the interface with the sea water was modified. The usual situation was that the groundwater increased its salinity degree during peak operation months, by even reaching drinking water standard limits.

The major evolution of artificial recharge in the recent decade does concern the nature and diversity of new compounds being the focus of new efforts leading to more and more advanced solutions of treatment. This is the result of the need to improve the water quality in line with the growing adoption of aquifers recharge for the use of potable reuse, especially for indirect application.

In the case of direct recharge, by means of water injection through deep wells (hundreds of meters depth), the water quality at the outlet of the treatment plant is compliant with drinking water standards (e.g., Australia, USA, Mexico). The coupling of "water infrastructure + direct artificial recharge" implies the integration of treatment technologies like the ones implemented for drinking water plants.

From the years 1960 to 1990, projects of indirect recharge did focus on the particulate and dissolved pollution in order to optimize and sustain the operability of the artificial recharge on the long term. The years 2000 have seen the increase of the artificial recharge for the consolidation and preservation of the water resource. The final purpose of the artificial recharge has evolved, now focusing more and more on the indirect potable reuse (and other usages) and on the aquifer storage. Treatment and monitoring have evolved toward guaranteeing advanced disinfection of treated wastewaters and micropollutant removal. The professional community is now focusing, through the configuration of different treatment lines, on the removal of antibiotic resistant bacteria (also called ARGs for AntibioResistant Genes) and microplastic removal.

Most recent infrastructure designed, built and operated by private process integrators demonstrate the increasing development of treated wastewater reuse coupled with direct artificial recharge, in terms of technologies but also in terms of regulation.

4.1 How to integrate MAR in water resource and treatment projects?

MAR utilizes and takes advantage of groundwater stores magnitudes, aquifer dynamics, physic-chemical processes as well as natural biological processes, in order to store and regulate surface resources, to fix problems derived from aquifer intensive exploitation, and even to purification water processes. Multiple solutions apply in the fields of hydrogeology and water treatment. MAR schemes and techniques are strongly dependent on local conditions.

Dams, treatment plants, canals, and water feeders are man-made water infrastructures that all engineers can see and imagine. They can also precisely design solutions. Groundwater and aquifers on the contrary are less visible, and not as well understood by the majority of water planners.

However, aquifers provide naturally and for free a high level of "environmental services" as water storage, water purification, water transport.

Even for a non-specialist, it is possible to include MAR in the design of many water projects, on the basis of three simple question marks:

- Case 1: Increasing Water storage (Dam or concrete storage). Is it possible to take advantage of natural groundwater storage? (Fig. 4)
- Case 2: Need for additional Water Treatment Plant (including desalination/brackish). Is it possible by aquifer recharge to benefit from better water quality or to restore naturally groundwater quality? (Fig. 5)

  A WTP could be replaced or its treatment CAPEX and OPEX minimized by appropriated MAR arrangement like River Bank Filtration.
- Case 3: Need for more Water Resource. Is it possible to increase local water resources by soil aquifer treatment, aquifer storage and reuse? (Fig. 6)

4.2 Treatment solutions to be considered for Aquifer Pre-treatment and Post-treatment applications

Like Pr. Michael E. Campana pointed out in recent publication, we need to "refill the hidden tanks" to "take all benefits of managing the recharge of aquifers."[10]

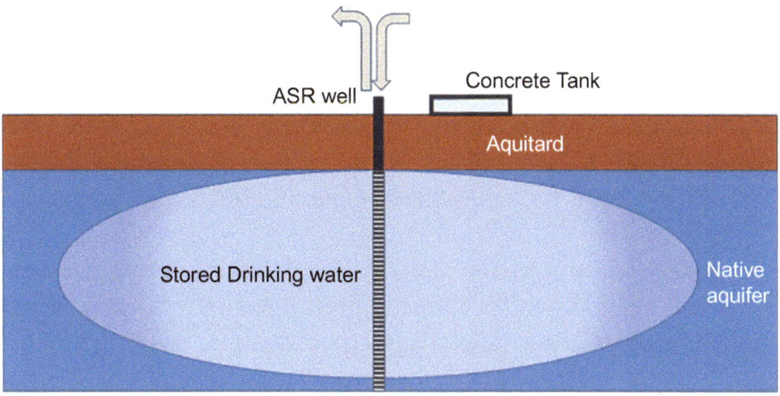

Fig. 4 Schematically representation of Drinking Water storage within underground aquifer compared to above-ground concrete or civil engineering tank storage alternatives.

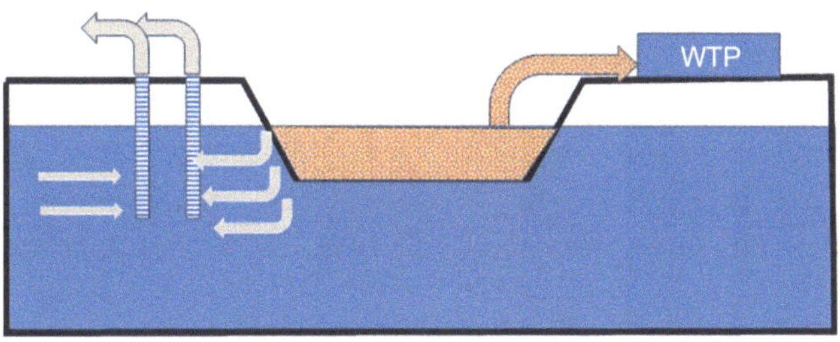

Fig. 5 Schematically representation of how soil aquifer treatment could improve water quality for aquifer withdrawal or prior to Water Treatment Plant processing.

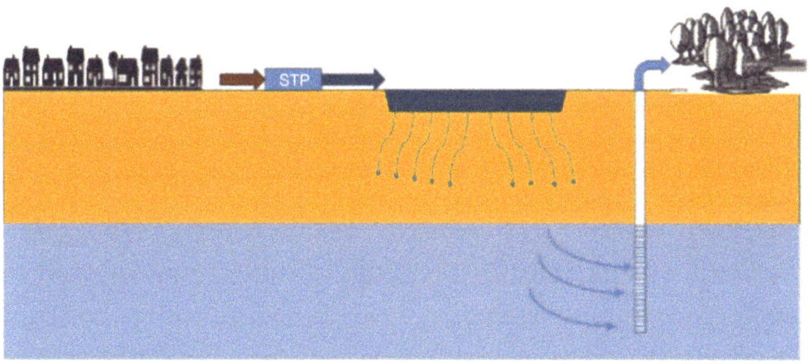

Fig. 6 Schematically representation of how MAR infiltration system and soil aquifer treatment can increase groundwater resource combined with downstream potable uses.

While considering aquifer recharge, water treatment through Pre- or Post-treatment is key to ensure safe and sustainable Water Reuse. When coupling Aquifer Recharge with Indirect Potable Reuse, benefits from storage and water quality monitoring arise and give credits to Water Reuse treatment solutions. In consequence pre- and post-treatment definition depend on the quality of the water resources, MAR processes features and end use purposes.

The aquifer recharge can be direct or indirect. If the aquifer is confined, recharge will be only possible through direct injection wells and therefore limiting the scope of soil natural treatment. In this case the quality of water injected must be, at least, better than native groundwater. Pre-treatment must then be more complex and expensive for these cases. In this case, the contribution of the aquifer as "Soil Aquifer Treatment" is minimized. When the aquifer is an unconfined one, it may be recharged indirectly, through infiltration basins, galleries or uncompleted wells thus permeating through unsaturated formations or soil. In these cases, soil acts as a biophysical reactor where water impurities are physically retained or transformed through bio energized redox processes, degassed and mitigated. Hence water pre-treatment before recharge could be simplified and "cheaper."

Each case is specific, and solution will vary depending on local regulation(s). However, treatment solutions can be summarized in three categories:

(a). Pre-treatment solutions for recharge of Confined Aquifer

In this case, recharge will only be possible through direct injection wells and the quality of water injected shall be at least better than the native groundwater. Pre-treatment will be more complex where Soil Aquifer Treatment is minimized. The quality of the water injected into the aquifer will fulfill the standards for drinking water. The water treatment process line and technological products will be the ones used by Suez for drinking water and desalination projects (Table 10).

Processes involved for confined aquifer and drinking water quality standards are the following:

(b). Pre-treatment solutions for recharge of Unconfined Aquifer

In this case, recharge may be indirect through infiltration basins, galleries or soils. In this case, soil acts as a biophysical reactor where water impurities are physically retained or transformed through bio energized redox processes, degassed and mitigated. Hence water pre-treatment before recharge could be simplified and cheaper (Table 11).

Table 10 Examples of Pre-treatment solutions for Confined aquifer.

Treatment solution	Suez degrémont™ solutions
Settling	Pulsatube™, PulsaGreen™, Pulsazur™
Dissolved air flotation (DAF)	AquaDAF™
Sand filtration	Aquazur™ V (sand)
Dual media filtration	Mediazur™
Filtration	Filter FH (DMPF), FV2B filters, Greenleaf™
Membrane ultrafiltration	Ultrazur™, Ultramarine™
Reverse osmosis (for micropollutants removal, for aquifers with brackish waters, sea intrusions)	Desalination membranes with pre-treatment such as SeaClean™, SeaDAF™, Cartridge filter, Ultramarine™
Activated carbon filtration (GAC, PAC)	Carbazur™, PulsaGreen™
Disinfection	Chlorination, Aquaray™ (UV)
Ozonation coupled with Biologically Active Filtration (disinfection, micropollutants)	Oxyblue™
Decarbonatation	Densadeg™
Mineralization	Lime saturation Densadeg™, Gyrazur™

Table 11 Examples of Pre-treatment solutions for Unconfined aquifer.

Treatment solution	Suez degrémont™ solutions
Advanced settling	Densadeg™
Dissolved air flotation (DAF)	GreenDAF™
Biofiltration	Biofor™
Micro-screening by mechanical sieve	Compakblue™
Sand filtration	Aquazur™ V (sand)
Dual media filtration	Mediazur™
Activated carbon filtration (GAC, PAC)	Carbazur™, PulsaGreen™
Membrane biological reactor (MBR)	Ultrafor™, UltraGreen™
Membrane ultrafiltration	Ultrablue™
Disinfection	Chlorination, Aquaray™ (UV)
Ozonation (disinfection, micropollutants)	Oxyblue™

Table 12 Aquifer Post-treatment solutions.

Treatment solution	Suez degrémont™ solutions
Iron removal	Ferazur™
Manganese removal	Mangazur™
Arsenic removal	Separation, chemical, biological or resin solutions
Heavy metals	Mining solutions

Processes involved for unconfined aquifer and high recycled water quality standards are the following:

The quality of the water injected into the aquifer will be closer to the highest standards applied for the treated water obtained by the most advanced treatment from waste water treatment plants. The water treatment process line and technology will be the ones used by Suez for advanced reuse projects including tertiary + quaternary treatment stages.

(c). The third category is the Post-treatment (combined or not with Pre-treatment)

In particular cases, when a specific pollution shall be treated within a polluted aquifer and where Soil Aquifer Treatment is minimized or not possible, the following technological product might be used for specific treatment (Table 12):

5. Pioneering References and Flagship Water Reuse Projects around the World

The growing needs in megacities for environmental integration with advanced odors management, retrofitting-extension of existing infrastructures and water reuse schemes tend to promote MBR as the most effective, efficient and sustainable solution. In certain regions with hydric stress, the development of water reuse has promoted the MBR technology because of its outstanding water quality, reliability and performance. Several MBR integration have been developed in past 5 years in the USA, China, France, Mexico within interesting water reuse flowsheet using MBR technology as core building block.

In the *West Texas City of Abilene (USA) the Hamby WRF Indirect Potable Reuse Project*[11] (12 MGD) has been in operation since January of 2015. The treatment flowsheet includes MBR technology, with 60% of the filtrate treated with reverse osmosis, and the other 40% treated through ozone oxidation and biologically active filtration.

The *Brightwater Wastewater Treatment Facility*[11] (117 MLD daily average flow to 217 MLD peak hourly flow) has been commissioned in September 2011. King County's Wastewater Treatment Division services about 1.4 million people in the Seattle, Washington area (USA).

The *Huai Fang Water Reclamation Plant*[11] project (located South of Beijing) has chosen the MBR technology (600 MLD; handling the wastewater produced by over 3 million people) equipped with the ultrafiltration-membrane bioreactor technology and ozone system used to treat the wastewater to produce optimum quality water, thereby contributing to the conservation of the natural environment receiving the outflow. At ground level, an artificial water lake uses reclaimed water for enhancing the natural environment. The effluent quality of the plant is designed to achieve "the urban sewage treatment plant water pollutant discharge standards of Beijing" (DB11/890-2012) requirements. The wastewater treated by the Huai Fang plant can be reused for city management projects or discharged into wetland and watercourse. This plant, that is built underground, improves the quality of the water environment and produce sound environmental benefits in Beijing southwest region.

The *Achères Biosav*[11] project (located downstream of the Paris' Seine river, one of the largest WWTP in the world; 250 MLD of UFMBR) has chosen to integrate the ultrafiltration MBR technology in hybrid configuration together with its existing secondary process (BAF). Designed in hybrid configuration (BAF + MBR) to improve the membranes' capital cost (CAPEX), the global energy efficiency (OPEX) and the final treated water quality. While treating 13–17% of the total flow (250,000 m^3/day @95%ile), the MBR treatment line guarantees 95%ile Total Suspended Solids (TSS) of 1 mg/L, Biological Oxygen Demand (BOD) of 5 ppm Total Nitrogen (TN) of 10 mg/L and Total Phosphorus (TP) of 0.3 mg/L. The combination of both technologies allows extending the treatment capacity of the plants within a limited footprint to improve and preserve the surface water receiving body. In addition, MBR technology provides an excellent water quality, which is used by ancillaries in the entire Achères plant as industrial water.

The *Chapultepec Mexico City MBR project*[12] (15 MLD; 105,000 PE) implements Indirect Potable Reuse scheme using UFMBR for multi-purpose water reuse. UFMBR is combined with UF-RO-UV for groundwater recharge.

The Chapultepec project is real symbol for the protection of water resource and citizen's welfare in the heart of Mexico City. Water scarcity is a reality in Mexico City. Around 75% of the 21 million inhabitants of the metropolitan area of Mexico City depend on groundwater resources. With the new Chapultepec WWTP project, the CONAGUA (the National Water Commission of Mexico) is tackling the water scarcity problem to its roots after having published in 2014 a comprehensive overview of the water sector in the country. In ancient times, before pre-Hispanic time (1325), the Aztecs established their capital Tenochtitlan (the former name of Mexico City) on an island. The city was surrounded by the giant lakes of Texcoco, Zumpango, Xochimilco, Chalco and Xaltocan. The Aztecs named the Chapultepec Hill the "grasshopper hill" of Tenochtitlan. When Hernan Cortes—the Spanish Conquistador—arrived in the 16th century (1521), the lake of the origins was magnificent. Before the Spanish settlement, "the Aztecs expanded the city with landfill and planted crops on floating gardens called chinampas, plots of arable soil created from wattle and sediment. The lakes provided the Aztecs with a line of defense, the chinampas with sustenance. The Aztecs managed to live with nature, but they were only 300,000 people. Then, the conquering Spaniards waged war against water, determined to subdue it. The Aztec system was foreign to them. They replaced the dikes and canals with streets and squares. They drained the lakes and cleared forestland, suffering flood after flood, including one that drowned the city for five straight years."[13] The route to urbanization begun and the rise of demography followed. The Aztecs had 300,000 people, today Mexico City counts 21 million. Up to the 16th century, there were five lakes in the valley covering more than 2000 km^2. Today, most of the lakes' area has been occupied by urban development with an extension close to 1900 km^2.

Nowadays, the lake of the origins does not exist anymore. It has been replaced by urbanization. Chapultepec is not the grasshopper hill anymore but remains the green lung of Mexico City. The development of the City of Mexico has gone through various regional processes and urban forms.[14] The lake city of the Aztecs originally served as a basis for urban concentration which resulted in the colonial city, which was marked by constant population growth and territorial expansion, intensifying during the 20th century.[14] In addition to this, the geographical and topographical features, availability of water resources and the location of centers of consumption and disposal, increase the complexity in the provision of potable water, sewage, treatment and reuse. On the other hand, the lakes were drained, forests were deforested, and water extracted from the aquifer more than its natural renewal capacity. This situation results in a growing need to ensure the supply of drinking water to the population, in quantity and quality; at the same time, it requires the timely evacuation of sewage and rainwater. Despite efforts to meet the most basic needs of water services, it has not been possible to meet these demands in full. The management of water resources is reaffirmed today more than ever to natural conditions in the Valley of Mexico, in an environment of greater uncertainty imposed by the possible impacts associated with climate change. Under these circumstances, today the priorities of SACMEX for the coming years are on services, resource preservation and water infrastructure.[14] Mexico City is considered one of the largest cities in the world and is home to 20% of Mexico's entire population. The estimated population of 8,918,653 in 2016 is up from 8.851 million in 2010. The metropolitan area, however, is much larger with a population of 21.2 million people, making Mexico City the most populous metropolitan area in the Western Hemisphere.[15] Around 75% of the 21 million inhabitants of the metropolitan area of Mexico City depend on groundwater resources.[16] Mexico is the 11th most populous nation in the world with an estimated 121,736,809 million people in 2016. Mexico's population is really growing in leaps and it's touted to soon overtake the population of Japan. When this happens, Mexico will be among the 10 most populous countries in the world today. A total of 79% people in Mexico reside in urban areas leaving a mere 21% to live in the rural areas. Roughly half of the population lives in 55 metropolitan towns. One of the reuse drivers is the Protection of the population. In general, the higher is the water stress, the lower is the water quality. This is true here. The aquifers of Mexico City have the lowest water level and the lowest water quality of the country.[17]

In 2014, the National Water Commission of Mexico (CONAGUA: Comisión Nacional del Agua) published a comprehensive overview of the water sector in the country emphasizing that aquifers and hydrometeorological phenomena are the main source of water supply in the North, North West and Center of Mexico (which represents two thirds of the country), because these regions are located on the fringe of the great deserts.[16] Across the country, the renewable amounts of water have been compared to the granted water for all uses, which revealed deficits in regions where aquifers are overexploited. Among the regions of the North, North West and Center, where the degree of pressure on water resources is significant, aquifers of Mexico City appears however to be the most under water stress.[17] In this context, reuse and aquifer recharge do represent a key scheme for sustainable water resource management. The CONAGUA promotes and maintains technical collaboration with other agencies to implement innovative measures for the protection and recharge of aquifers.[16] Proof of this is the strategy of infiltration in the aquifer of the Mexico Valley, projects in San Luis Rio Colorado, Sonora and Tabalaopa-Aldama, Chihuahua.[16] Furthermore, the 2014-study highlighted that the very high-water scarcity degree in Mexico City was correlated to lower water quality in this region. BOD_5, COD and TSS concentration measurements were the highest of the country in the Valley of Mexico.[16]

Considering the water stress in the Valley of Mexico, the critical water level of the aquifer under Mexico and the need for enhancing water reuse in the country, Indirect Potable Reuse and Direct Potable Reuse are strategic schemes for sustainable development in the region. The resource preservation in the heart of Mexico City is essential as the Chapultepec Park and Forest remains one of the last "green lung," which ensures water resort for Mexicans, recreational attractiveness, cultural (museum) and economic activities: 678 ha, 3 lakes, 105 species of trees, 222 species of birds and mammals. The project aims at preserving what is one of the largest urban parks in the world and the main green area of Mexico City. Chapultepec forest is not the "grasshopper hill" anymore, but the last "green lung of the City that must be preserved." The Federal Government has made the preservation of Chapultepec Park a key stake with the local authorities involved in the Comprehensive Rehabilitation Program such as the National Water Commission through the General Department of Water Supply, Sanitation and Drainage. The need for improving water infrastructure has been scheduled by the authorities to allow preserving the Chapultepec forest, which remains an important recreational space for the Mexicans. The current infrastructure for the treated water distribution will be used and optimized within the site. Building a new wastewater treatment plant infrastructure allows producing the flow of treated water more than necessary to meet the requirements of the Chapultepec Forest.

Indirect reuse will be implemented for water filling of the lakes with recreational vocation and irrigation of green areas. The surplus of treated water not used for the lakes will be further treated for polishing treatment to meet the quality requirements for ensuring injection in the underlying aquifer. Groundwater recharge will be implemented to reinforce the water table level of the aquifers of Mexico City. The project integrates a three-stage membrane barrier (MBR(UF) + UF + RO) with regards to the need for groundwater recharge (aquifer refilling) and a UF filtration + UV Disinfection for the water supply to lakes and irrigation of the forest. The treated water from the plant will contribute to restore the Chapultepec's lakes aesthetic appearance (from green to blue) and the utilization of the sludge produced by the WWTP will be, after lime stabilization, used as fertilizer in the soils of the Chapultepec Park. The polishing plant dedicated to aquifer recharge and composed of Pre-treatment-UF-RO will have a capacity of 6900 m^3/day. These technologies will allow complying with the water quality requirements needed for a direct recharge of the aquifer.

The National Water Commission of Mexico has combined with this Water Reclamation Plant (WRP) the actual need to improve recreational activities within Chapultepec Park for the sustainable development of the city and has made a significant step forward in contributing to reinforce the water resource in Mexico City. IPR is a key scheme for sustainable development in the region. Considering the need for increasing the quantity and for improving the quality of its water resource, it is expected that Mexico will increase its water reuse rate in light of a comprehensive water management, such as to focus much more on IPR and DPR in the near future, when compared to the current share of Agriculture. With indirect water reuse and sludge reutilization both within the Chapultepec Park, this makes the Chapultepec WRP the most emblematic example of circular economy by means of advanced technologies coupled with resources recycling (Fig. 7).

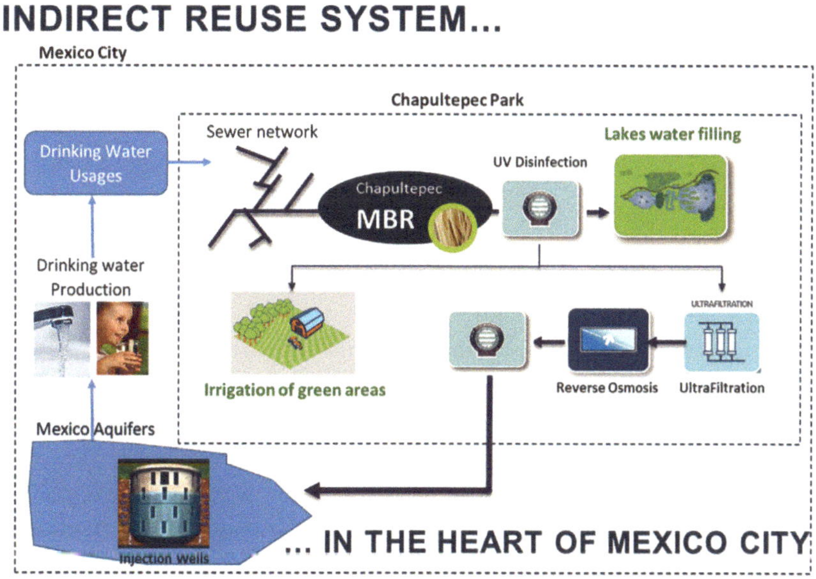

Fig. 7 Indirect Potable Reuse Scheme in the context of the Chapultepec project.

Table 13 Mexican standard for Aquifer replenishment NOM 014-Conagua 2003. **Direct injection: artificial recharge**

Pathogens microorganisms	Removal or total inactivation of whole pathogens
Contaminants regulated by drinking water standards	Limits by NOM-127-SSA1-1994
Contaminants not regulated by drinking water standards	COT $\leq$ 1 mg/L

Table 13 summarizes the water quality levels for aquifer replenishment in the country of Mexico, also used for the artificial recharge implementation for the Chapultepec project.

The Edward C. Little Water Recycling Facility in El Segundo West Basin facility (USA, South California) is a real pioneer and unique infrastructure developed by the Municipality since no less than 25 years from 1995. West Basin Municipal Water District (WBMWD) is a Public Agency responsible for drinking water wholesale and wastewater recycling for the West part of Los Angeles, California. The Agency has developed since 1995 a very ambitious reclamation program, focused around the Edward C. Little Water Recycling Facility (ECLWRF). Suez group has accompanied the Agency's growth ever since the beginning of their program and operated the ECLWRF and its satellites facilities since 1995 under 5 years Operation and Maintenance services contracts.

The West Basin Edouard C. Little facility implements a Multi-purpose Water Reuse Supply with a dedicated Tariffs function of final usages. In Southern California West Basin, the Edward C. Little Water Reclamation Plant (107 MLD) does gather the entire panoply of recycled water possibilities. Five different treatment lines supply high standard water quality with four specific water tariffs and several water users off-takers. The plant produces five types of reuse water: seawater intrusion barrier water for groundwater recharge and indirect potable water (27%), nitrified water for industries refineries (26%), water for irrigation/urban non-potable uses (27%), ultrapure water for industrial process water (20%). California is dependent on imported water limited by infrastructure originally designed for 18 million people while the region serves 37 million in 2012. The water system is worsened by demographic pressure (48 million inhabitants in 2025: 140% increase since 1970), which generates a demand greater than offer, delta smelt restrictions and unstable delta levees. Drought and competing demands on water supplies emphasize the need for water supplies to create a more sustainable region. The facility provides water to industrials and municipal end-users and stakeholders, with a mix of industrial, commercial and Public Agency customers: Recycling Facility and three Refinery Satellite Operations (Chevron, Exxon Mobil and Carson's Tesoro), US Air Force, Marriott hotel and resorts, Toyota, Honda… with more than 300 customers. West Basin plant does contribute to increase water recycling from 7% to 22% and to reduce imported water from 66% to 33%. The plant is compliant with the Title 22 California Code of Regulations (Table 14).

The West Coast Basin Barrier Project (WCBBP) in California, initiated in the 1960s, consists of approximately 150 injection wells located along the coast. In 1995, the injection of 100% potable water was replaced by a blend of surface water and recycled water, with gradual increase of the proportion of recycled water from 35% to 100%. The high-quality recycled water is produced by the Edward C. Little Water Recycling Facility in El Segundo, which is world renown as the only facility producing five distinct types of "designer" recycled water. Thereby, treatment and cost are customized to the required use: tertiary disinfected water (Title 22 water quality) for irrigation and other urban uses and to feed satellite plants at three refineries, nitrified water for cooling towers, reverse osmosis followed by advanced oxidation water for direct aquifer recharge as a salt intrusion barrier, and single and double pass reverse osmosis for low- and high-pressure boiler water.

This unique concept is a model to replicate as it enables successful implementation of integrated resource management in urban areas with tailored solutions integrating creative project funding with the issuance of bonds and grants and public-private partnership for operation. It followed a step by step expansion with implementation of innovative tools and technologies, integrated management of aquifer recharge: groundwater models show the injected water has a 10 to 20-year residence time in the ground prior to being pumped out in groundwater wells and used as a potable source.

Table 14 West basin recycled water rates.[18]

Recycled Water usage	**Tariff in USD/m^3**
Low-pressure boiler (refinery industry)	1.08
High-pressure boiler (refinery industry)	1.37
Cooling tower (industry)	0.88
Groundwater recharge (municipal)	0.68

In the end, benefits of such exemplary model are multiple: economic benefits for the population due to reliable water supply at moderated cost, increase in land and property value and positive impacts for the human wellbeing, sport and tourism, economic benefits for industry coming from avoided economic losses during drought periods and loss of a key industries to the region, social benefits of job saving, as tens of thousands of jobs would have been lost to the region without the addition of recycled water to supplement the supply of potable water, environmental benefits for the improvement of groundwater quality, efficient control of seawater intrusion and preservation of the precious local groundwater resources. The ECLWRF uses a small portion of the treated water from the Hyperion WWTP. This wastewater treatment facility (carbon removal, 2020) is the largest so far in California (1700 MLD Dry Weather Flow; 3200 MLD Wet Weather Flow).

The Perth Beenyup's facility (Western Australia) is integrated in the Perth Groundwater Replenishment scheme. For almost 2 million people living in the Perth metropolitan area, 335 million liters of wastewater are produced every day and released mostly to the ocean following treatment. Currently, only 6% of wastewater is recycled, although the Water Corporation aims to increase this figure to 30% by 2030.[19] Groundwater replenishment (GWR) using secondary treated wastewater flows from the Beenyup Wastewater Treatment Plant (WWTP) was identified as a preferred water source option in the Water Corporation's strategic planning document: Water Forever—toward climate resilience. Subsequently, GWR has been identified as a critical component of the Groundwater Security Strategy (GWSS), the Corporation's 10 years plan to maintain groundwater as the primary water source for the northern part of Perth.[19] The Water Corporation undertook a Groundwater Replenishment Trial (GWRT) at the Beenyup WWTP site from 2010 through to 2012 to assess the technical and social feasibility of an advanced water recycling process for producing water that is suitable for replenishing groundwater. The Trial was operated in accordance with the Australian Guidelines for Water Recycling and successfully demonstrated a treatment train.[19] In the end of 2014, the construction of the first Australia's full-scale groundwater replenishment scheme started with the first stage of the Advanced Water Recycling Plant (AWRP). The first stage (14 GL expected production per year—capturing almost 50% of Beenyup treated wastewater) was planned for being in operation in December 2016 by the Aroona Alliance (Alliance between Water Corporation, SUEZ and Broadspectrum). The second stage was implemented from 2017 to 2019.

Project drivers include: water demand exceeding local water supply (groundwater capacity and surface water); over abstraction to local aquifers, strategy to become climate resilient/a few years of acute climate drought; solution less expensive than desalination (including conveyance costs), strong support from the Western Australia State government, successful pilot stage undertaken in 2010–2012 and full-scale infrastructure project finalized from 2017 to 2019.

This project includes: four treatment steps: UF, RO, UV disinfection and injection in deep aquifers; a maximum capacity of 44,000 m^3/day, injection in groundwater of 14 million m^3 per year in the first stage through four recharging bores; abstraction from the same aquifers to produce potable water in equivalent quantities (Table 15).

This project has economic benefits for the population due to a climate resilient water supply source, a more competitive cost than desalination (no conveyance is required by using the large Perth aquifers); environmental benefit by controlling the water balance for the deeper groundwater resources. Overall, this project contributes to limit the abstraction from superficial aquifers and allows social benefit by enabling the Perth metropolitan growth and limiting the impact of the recurrent climate drought that do affect Western Australia.

Even on the East coast of the United States, where water scarcity does not make the headlines of the news, Wastewater Reuse projects develop.

Table 15 Microbial log reduction credits in the context of Perth Beenyup implementation.[19]

	Equivalent log reduction credits		
	Bacteria	**Virus**	**Protozoa**
Wastewater treatment plant	1	1	0.5
Beenyup advanced water reclamation plant			
UF with chloramination > 1.5 mg/L	3	3	3
Reverse osmosis	3	3	3
UV disinfection > 186 mL/cm^2	4	4	4
Total AWRP	10	10	10
Total WWTP + AWRP	*11*	*11*	*10.5*
Department of Health Requirement	*8.5*	*9.5*	*8*
Excess credits (safety margin)	2.5	1.5	2.5

This is the case of *the Hampton Roads Sanitation District (HRSD) Groundwater Injection Project* (Virginia). The Virginia Coastal Plain faces groundwater supply depletion, land subsidence, saltwater intrusion from the Potomac River down to the Chesapeake Bay. HRSD has targeted by 2030 to recharge the groundwater supply by treating wastewater to drinking water quality standards and injecting it into the aquifer.[20] The SWIFT project aims at addressing sea level rise: "*taking water out of the ground at the current rate has led to sinking of land, or land subsidence, in some parts of eastern Virginia. This makes us more vulnerable to rising sea levels and the associated impacts*" and helping the Chesapeake Bay "*restoring the health and productivity of the Chesapeake Bay largely depends on reducing the amount of nutrients and sediment that enter Chesapeake Bay waters.*" "*By 2030, HRSD's SWIFT Water is projected to effectively eliminate more than 90% of HRSD's discharge to local waters. This will reduce the total amount of nutrients, such as phosphorous and nitrogen, reaching the receiving streams, and ultimately the Bay.*" In this project HRSD implemented advanced pilot testing of different treatment flowsheets to meet drinking water standard from wastewater reuse before aquifer injection. All membrane trains and ozone/carbon-based solutions have been tested, such as MF-RO-UVAOP, MF-NF-UVAOP and FLOC/SED/OZONE-BAC-GAC-UVD.[21]

In South California, *the DC Tillman Advanced Water Purification project*[22] *of Los Angeles Sanitation* issued a report in July 2018, studying treatment performances and Life-Cycle Cost Analysis of ozone-based advanced treatment "Tertiary Treatment-Ozone-BAC-LPUV," membrane based treatment "Tertiary Treatment-MF-RO-UV/AOP" and hybrid blended treatment trains "tertiary treatment-Ozone-BAC-sidestream MF-RO followed by mainstream LPUV."

The report specifies "*the recommended AWPF treatment train for the full-scale AWPF system is the ozone-based treatment train (O3-BAC-UV-SAT) (...) along with simultaneous development of an alternative to TOCmax as a framework for public health protection that is responsive to the California Code of Regulations (...) Sidestream treatment with MF and RO or granular activated carbon (GAC) for a small portion of the water is another strategy that could be employed as a backup option for additional TOC or salinity removal, if necessary. The ozone-based treatment train offers the following key benefits:*

- *Higher recovery and production of purified water (24.5 mgd vs 16.8 mgd)*
- *Lower energy consumption*
- *Reduced waste (no waste stream produced for treatment and management)*
- *Lower capital and operating expenses, resulting in lower costs to LADWP customers*"

The DC Tillman is the kind of leading-edge project and new flagship advanced treatment facility that will symbolize the new trends occurring in California Water Reuse when it comes to finding alternative to all membrane trains. This project will surely become a cornerstone that will influence new trends in the development of wastewater reuse in the USA and worldwide in the next two decades, from 2020 onward to 2040.

6. Horizon 2040 of water reuse

6.1 Indirect potable reuse coupling wastewater reuse with managed aquifer recharge

Aquifers are more than ever the nexus of Indirect Potable Reuse to increase global Water Resource, recharge local water cycle, centralize water monitoring and provide sustainable solution in many regions. As water resources and energy became scarcer in many regions of the world, aggravated by the consequences of Climate Change and demography evolution, Water Reuse combined with Managed Aquifer Recharge is a remedy, that provides concrete solution, combining quantitative and qualitative benefits. It secures and enhances water supplied at low cost by meeting seasonal peak demands thanks to aquifer storage. It reduces evaporation of stored water while reducing costly investments in traditional concrete water tanks. It improves water quality and prevents salt water from intruding into coastal aquifers. Aquifers constitute the most sustainable environmental buffer for an indirect potable water reuse scheme. Environmental benefits are numerous, such as maintaining flows and groundwater-dependent ecosystems, which improves local amenity, land value and biodiversity. It improves coastal water quality by reducing urban discharges, it mitigates floods and flood damage and facilitates urban landscape improvements that increase land values.

Dams, treatment plants, canals and water feeders are man-made infrastructures. Groundwater and aquifers on the contrary are less visible and not well understood by most water planners. However, aquifers provide naturally a high level of "environmental services" such as: storage, purification and transport. Management of Aquifer Recharge should be included in the design of water projects, based on simple drivers and three pragmatic points:

- **i.** Increasing water storage (Dam or concrete storage) by considering an advantage to natural groundwater storage
- **ii.** Need for additional Water Treatment Plant (including desalination) by taking advantage from aquifer recharge to benefit from better water quality or to restore naturally groundwater quality

iii. Need more Water Resources to increase local water resources by reuse, soil aquifer treatment and aquifer storage

Geologist and water treatment engineers are committed worldwide to design and provide fit-for-purpose water reuse solutions including the sustainable approach of Management of Aquifer Recharge.

6.2 IPR and the rise of Direct Potable Reuse

If the first Direct Potable Reuse (DPR) project was developed and implemented in Namibia for the city of Windhoek, Indirect Potable Reuse (IPR) and DPR remain applications led by USA and South California in particular. Australia follows closely with key applications like in Perth Beenyup (2017), designed and built by Suez for the Water Corporation is the first IPR facility in Australia.

Back in time, the Water Factory 21 of the Orange County Sanitation District (OCSD) is the pioneer and the oldest world reference in wastewater reuse, back in 1976. West Basin Edouard C. Little Facility is one them since 1993. Big Spring (Texas) in 2013 and Wichita Falls (Texas) in 2014 are two of the most recent DPR facilities. More recently in 2017 Terminal Island Advanced Water Purification Facility (CA) implemented IPR (Fig. 8). In 2017, US EPA released the Potable Reuse Compendium highlighting that as of 2017, there are about 50 Water Reuse projects planned in USA (IPR and DPR). On the West coast (California, Washington, Oregon), South (Texas) as well East coast (Florida, Georgie, Tennessee, Virginia).

From 2020 onward, two infrastructure projects perfectly embody the new trend of wastewater reuse in years to come: *the demonstration Facility of Sanitation District of Los Angeles County* that aims validating the ability of MBR-RO-UV/AOP to meet the DPR-IPR investment program objectives of Municipal Water District of Southern California; *the future DC Tillman Advanced Water Purification facility of Los Angeles Sanitation* is the other key flagship project in South California and beyond.

6.3 Agriculture

In 2004, "Water Reuse for Irrigation: Agriculture, Landscapes, and Turf grass"[24] was published specifying water reuse treatment qualities objectives for fit-for-purpose agriculture (Table 16).

In "2012 Guidelines for Water Reuse" from the US EPA,[25] recommendations for agricultural irrigation mention that "*no detrimental effects are usually noticed below 500 mg/L. Between 500 and 1,000 mg/L, TDS in irrigation*

Year	Milestone Facilities
1962	**Montebello Forebay Spreading Grounds, Los Angeles, CA**
1968	Goreangab **DPR Plant I, Windhoek Namibia**
1976	**Water Factory 21, OCWD, CA**
1978	Occoquan Reservoir, **UOSA, VA**
1985	Hueco Bolson Recharge Project, **El Paso, TX**
1993	**West Basin Water Recycling Facility**, Carson, CA
1999	Gwinnett County, GA
1999	Scottsdale Water Campus, AZ
2003	NeWater Bedok, Kranji, **Singapore**
2005	Water Replenishment District, CA
2007	Chino Basin Recharge Project, CA
2008	Groundwater Replenishment System, CA
2009	Arapahoe & Cottonwood, CO
2010	NeWater, Changi, **Singapore**; Prairie Water, CO
2013	**Big Spring, TX (DPR)**
2014	Wichita Falls, **TX (DPR)**
2017	Terminal Island Advanced Water Purification Facility, **CA (IPR)**

Fig. 8 History of potable reuse supply 1962–2017.[23]

Table 16 Sequential irrigation of increasing salt tolerant crops with drainage water from less tolerant crops.[24]

	Type of crop				
Parameters	**Sensitive (peas, beans, strawberries, stone, pome, and citrus fruits)**	**Moderately sensitive (lettuce, kale, broccoli, celery, potato)**	**Tolerant (wheat, sorghum, rye, beet)**	**Very tolerant (barley, cotton, sugar beet, bermuda grass, salt cedar, eucalyptus, poplar)**	**Halophyte (salicornia)**
Irrigation volume[a]	100	25	10	5	2
TDS, mg/L	*200*	*800*	*2000*	*4000*	*10,000*
Efficiency, %	75	60	50	60	67
Drainage volume[a]	25	10	5	2	0.67
TDS, mg/L	*800*	*2000*	*4000*	*10,000*	*30,000*

[a]Volumes expressed in arbitrary units.

water can affect sensitive plants; at concentrations above 1,000–2,000 mg/L, TDS levels can affect many crops (...) At TDS concentrations greater than 2,000 mg/L, water can be used regularly only for salt-tolerant plants on highly permeable soils. (...) With respect to chlorine residuals, which may be present as a disinfection residual, free chlorine at concentrations less than 1 mg/L usually poses no problem to plants; chlorine at concentrations greater than 5 mg/L can cause severe damage to most plants. However, some sensitive crops may be damaged at levels as low as 0.05 mg/L. (...) To avoid leaf-burning effect avoid reclaimed water to be sprayed directly onto foliage."

Unfortunately, business reality shows that in the Municipal Wastewater Reuse sector, there is still a lack of precise water quality prescription in most conventional bid tender procedures for wastewater reuse projects targeting irrigation (agriculture and turf grass).

Wastewater reuse in agriculture is key for the future of water resource planning. As of 2020, there is a growing need for more specific prescription and more project development promoting fit-for-purpose water quality for agriculture by carefully anticipating the final use of recycled water in function of specific crops and plants.

6.4 Micropollutants Removal to Protect environmental bodies and fauna

Market applications of micropollutants removal in Wastewater treatment increase across Europe (Switzerland, Scandinavian countries, Germany, Spain, France). These applications of "advanced tertiary" will become by 2040 a significant market size for providing "superior" discharge water quality to the environment and for the protection of aquatic lives, especially in western developed countries. In North America (USA) the increasing concern around the wise spread of PFAS, PFOS and PFOA (specific but ubiquitously micropollutants) will more than likely generate upgrade of Drinking Water Plants and WWTPs to contribute to the capture and reduction of those compounds.

Growing concerns have emerged in the past 15 years about impacts of Micropollutants or Compounds of Emerging Concern (CEC) present in water. This being the result of the ever-increasing development and uses of mainstream medication worldwide, combined with the upsetting production of thousands new molecules by the chemical and pharmaceutical industries. Such concern and interests has quite recently allowed highlighting long term and unexpected changes of live stocks due to potential cocktail effects, endocrine disruptors, bioaccumulation, biomagnification, antibiotic resistance... In the past, detecting extraordinary event was implemented through the angle of mortality rates, such as the famous trout monitoring tests applied in Drinking Water Plant. In more recent years, ecotoxicological tests contributed to improve awareness with regards to long term and cocktail effects of emerging contaminants on aquatic wildlife. Fish tests have been developed and demonstrated impacts of different water qualities on rainbow trout early life development. Fish early life stage-tests support awareness of detrimental micropollution effects on lives. It also supports the efficiency of ozone and carbon solutions when it comes to lower toxicity effects and to recover the highest water quality level, as high as natural origin water.[26]

In continental Europe like in the eastern part of the United States, Reuse *De facto* is a reality. In EU for instance, surface water masses have really been improved over the past 30 years in terms of advanced removal of carbon and nutrients, thanks to important and continuous strategic investments on wastewater treatment infrastructure. Treated water discharged to lakes and rivers by WWTPs have dramatically improved as well as surface waters, especially in terms of particles and dissolved parameters. However, while existing WWTPs remove a minor portion of Micropollutants, emerging contaminants are still transferred to receiving bodies if no additional nor specific advanced treatment is implemented. While a majority of DWPs are equipped with adapted technologies to cope with those contaminants to protect human life, this is not the case for the protection of aquatic livestocks living in surface waters.

Comprehensive removal of micropollutants in the water cycle system can be carried out at WWTP discharge level, at DWP and through dedicated Potable Water Reuse infrastructure (direct or indirect), such as indirect potable reuse (IPR) scheme. When it comes to foster Micropollutants removal investments for the protection of water receiving bodies and fauna, or through the development of indirect potable reuse (IPR), awareness of decision makers and public outreach will be key in near future to emphasize the compelling need to engage broad investment efforts in fit-for-purpose micropollutants treatment at WWTP.

6.5 Discussion about Water Reuse in Europe

If European countries in Southern Europe have been particularly active in the field of water reuse activities (Spain, Italy), Europe remains far behind compared to USA, Australia (to quote only those). As the world leading experts are developing advanced water reuse all around the world, European Union has only recently (2020) encouraged Water Reuse for industry, agriculture and landscape irrigation. However, Indirect Potable Reuse through Managed Aquifer Recharge has been left behind. This happens when the adoption of IPR and the effects of Climate Change on water management are accelerating worldwide. Leading edge regulations in California and Australia have demonstrated that potable water reuse infrastructure projects help planning safe water resource in terms of quantity and quality (by also taking into account the most recent micropollutants). The development of IPR schemes through Managed Aquifer Recharge is key to anticipate the effects of climate change worldwide, and in Europe, from 2020 onward to 2040.

Acknowledgments

I want to thank my colleagues from Suez International who are committed to provide the best water treatment solutions in many important projects for government and population around the world: Miguel-Angel Sanz, Sylvie Baig, Philippe Gislette, Valentina Lazarova, Stephen Katz, Alejandro Zuluaga, Angel Canales, Jéromine Albertini, Antoine Vuillermet, Adriana Gonzalez-Ospina, Jean-Michel Grenaingaire, Fabian Fenoglio, Laurent Guey, Chrystelle Langlais.

Thanks to Raphael Mujeriego, Jörg Drewes, Greg Wetterau, Gil Crozes, Domingo Zarzo, Akissa Bahri for the inspiration they have transmitted during numerous international congresses I could participate.

I want to address a very special thanks and my recognition to Miguel-Angel Sanz, Past President of the International Desalination (and Reuse) Association (IDA), from whom I am very grateful to have given me the opportunity since 2015, to develop my professional career and my domain of expertise from wastewater treatment to water reuse.

References

1. University of Maryland, ScienceDaily. *Major shifts in global freshwater, first of its kind study combines NASA satellite observations of Earth with data on human activities to map where and why freshwater is changing around the globe.* https://www.sciencedaily.com/releases/2018/05/180516162536.htm; May 16, 2018.
2. NASA. *Global map of freshwater stored on land.* Gravity Recovery and Climate Experiment (GRACE); February 2016. https://gracefo.jpl.nasa.gov/news/131/nasa-satellites-reveal-major-shifts-in-global-freshwater/.
3. Figure SPM.1, From IPCC. *Summary for policymakers.* In: Pachauri RK, Meyer LA, editors. *Climate change 2014: synthesis report. Contribution of working groups I, II and III to the fifth assessment report of the intergovernmental panel on climate change.* Geneva, Switzerland: IPCC; 2014 Core Writing Team. https://www.ipcc.ch/site/assets/uploads/2018/02/AR5_SYR_FINAL_SPM.pdf.
4. Suez, Donnaz S, Canales A, Zuluaga A, Mauro E, Tieche A, et al. *Protection of water resource and citizens' welfare in the heart of Mexico City: Chapultepec WWTP—58075Donnaz.* São Paulo, Brazil: The International Desalination Association World Congress; October 2017.
5. Smith CDM, David Prah PE, BCEE. *Water purification technology evaluation and R&D testing project, 2019 excellence in environment engineering and science award winner, honor award in*

the category research. American Academy of Environmental Engineers & Scientists; 2019. https://www.aaees.org/e3scompetition/2019honor-research.php.
6. EPA United States, Environmental Protection Agency. *Potable reuse compendium chapter 7-6, table 7-2*. https://www.epa.gov/sites/production/files/2018-01/documents/potablereusecompendium_3.pdf; 2017.
7. EPA United States, Environmental Protection Agency. *Potable reuse compendium chapter 6-17, table 6-3*. https://www.epa.gov/sites/production/files/2018-01/documents/potablereusecompendium_3.pdf; 2017.
8. WHO. *Guidelines for drinking-water quality third edition incorporating the first and second addenda, volume 1, recommendations*. https://www.who.int/water_sanitation_health/dwq/fulltext.pdf; 2008.
9. Suez, Donnaz S, Sanz M-A, Baig S, Gislette P. *IWA Reuse Longbeach 2017, IWA Reuse Berlin, 2019, "Recharge local water cycle from surface water and wastewater reuse"* 2019.
10. Oregon State University, Michael E. Campana, Professor of Hydrogeology and Water Resources Management at Oregon State University [the 'other OSU'] in Corvallis, OR and Technical Director of the American Water Resources Association, IWA website, https://www.thesourcemagazine.org/refill-hidden-tank-benefits-managing-recharge-aquifers/, "Refill the hidden tank: the benefits of managing the recharge of aquifers", October 2016
11. Suez, Donnaz S, Katz S, Sanz M-A. *MBR—The corner stone of water reuse adoption*. IWA Water Reuse Berlin; 2019.
12. Suez, Donnaz S, Canales A, Zuluaga A, Sanz M-A. *Indirect water reuse in the heart of Mexico City: Chapultepec WWTP*. IDA Reuse Nice; 2016.
13. The New York Times, Kimmelman M. *"Mexico City, parched and sinking, faces a water crisis", climate change is threatening to push a crowed capital toward a breaking point*. https://www.nytimes.com/interactive/2017/02/17/world/americas/mexico-city-sinking.html; February 17, 2017.
14. *Programa de Gestión Integral de los Recursos Hídricos, Visión 20 Años*. SACMEX; October 2012. http://www.agua.unam.mx/sacmex/assets/docs/PGIRH_Final.pdf.
15. Mexico Population. *World population review*. https://worldpopulationreview.com/countries/mexico-population; 2017.
16. Suez, Donnaz S, Canales A, Zuluaga A, Mauro E, Tieche A, et al. Sustainable indirect water reuse in Mexico City: advanced treatment using membranes, In: *11th International Water Association international conference on water reclamation and reuse, Long Beach, California, July 23–27*; 2017.
17. CONAGUA. *Estadísticas del Agua en México edición 2014*. Comisión Nacional del Agua (CONAGUA); 2014. http://www.conagua.gob.mx/CONAGUA07/Publicaciones/Publicaciones/EAM2014.pdf.
18. West Basin Municipal Water District. *FY2017–18 water rates and charges, recycled water rates effective*. https://www.westbasin.org/finance/water-rates-charges; July 1, 2017.
19. Water corporation of Western Australia. The principal supplier of water and wastewater services to over 2 million people across Western Australia, In: *The International Desalination Association conference on water reuse and recycling—Nice, France*; September 2016.
20. LA Sanitation, City of Los Angeles, Brown and Caldwell, Trussell Technologies Inc. *Advanced water purification facility pilot study report Donald C. Tillman Los Angeles groundwater replenishment project*. https://www.lacitysan.org/san/sandocview?docname=cnt035139; July 16, 2018.
21. *The SWIFT HRSD groundwater injection project at*. www.hrsd.com/swift.
22. Nading T, Schimmoller L, Holloway D, Henifin T, Dano J, Salazar-Benites G, et al. *A SWIFT approach to managed aquifer recharge*. https://www.wateronline.com/doc/a-swift-approach-to-managed-aquifer-recharge-0001 Water Online; January 24, 2018.

23. Carollo Engineers Inc, US, Gil Crozes. The state of potable reuse in Southern California, In: *The International Desalination Association conference on water reuse and recycling—Valencia, Spain*; June 2018.
24. CRC Press, Lazarova V, Bahri A. *Water reuse for irrigation, agriculture, landscapes, and turf grass.* https://fr.scribd.com/document/425362469/Water-Reuse-for-Irrigation; December 28, 2004.
25. EPA United States, Environmental Protection Agency. *2012 Guidelines for water reuse, 3.2.2.1 salinity and chlorine residual, 3–8.* https://www.epa.gov/sites/production/files/2019-08/documents/2012-guidelines-water-reuse.pdf; September 2012.
26. Swiss Centre for Applied Ecotoxicology, Eawag/EPFL, Kienle C, Kase R, Werner I. *Evaluation of the ecotoxicity of sewage plant wastewater from WWTP basel with bioassays before and after the use of advanced wastewater treatment methods.* Final report March 31st. https://www.centreecotox.ch/media/2229/2011_kienle_bioassays_micropoll.pdf; 2013.

CHAPTER FOUR

Impact of the use of treated wastewater for agricultural need: Behavior of organic micropollutants in soil, transfer to crops, and related risks

Lapo Renai[a], Francesca Tozzi[b], Leonardo Checchini[a], and Massimo Del Bubba[a,*]

[a]Department of Chemistry, University of Florence, Florence, Italy
[b]Council for Agricultural Research and Economics, Research Centre for Vegetable and Ornamental Crops, Pistoia, Italy
*Corresponding author: e-mail address: massimo.delbubba@unifi.it

Contents

1. Introduction 104
2. Use of WWs in agricultural production: Aim and consequences 106
   2.1 Highlights on the use of WWs for irrigation purposes 106
   2.2 OMPs in WWs used for agricultural irrigation 117
3. Behavior of OMPs in soil 118
   3.1 Highlights on OMPs occurrence in soil 118
   3.2 Influence of soil properties on OMPs sorption/desorption and degradation processes 119
4. Transfer of organic micropollutants in edible crops 121
   4.1 Mechanisms of plant uptake and translocation of OMPs 121
   4.2 Occurrence of OMPs in edible crops 123
   4.3 Assessment of the risk associated with the intake of food contaminated with OMPs 124
5. Concluding remarks 128
References 129

Abstract

Agriculture represents a sector typically characterized by a high water demand. On the other hand, limited fresh water availability is a problem of increasing concern worldwide. The reuse of treated wastewater (TWW) for edible crop irrigation could be an efficient strategy for reducing water shortage. However, TWW reuse poses risks for both environment and human health due to the transfer of chemical contamination from water to soil and edible crops. This chapter discusses the behavior of various categories

Advances in Chemical Pollution, Environmental Management and Protection, Volume 6
ISSN 2468-9289
https://doi.org/10.1016/bs.apmp.2020.07.008

of organic micropollutants (OMPs) in the wastewater/soil/edible crop systems, with particular reference of pharmaceutical compounds (PhCs). Moreover, the assessment of the risk associated with the intake of several fruits and vegetables contaminated with OMPs was performed for some model organic contaminants widely investigated in literature. Data were reviewed using the search engines Google Scholar and SciFinder, and referred to the time intervals from 2010 or 2005 to today for PhCs and other OMPs, respectively.

Keywords: Pharmaceuticals, Personal care products, Soil sorption, Degradation, Plant absorption and uptake, Dietary exposure

1. Introduction

Agriculture is facing challenges related to a large demographical explosion, which in 2050 should lead the world population to exceed 9 billion people. Furthermore, agriculture itself is already and will be still more affected by climate change, which is likely to cause a reduction in natural resources, in particular fresh water (FW) and fertile soil, as well as significant changes in plant growth conditions, such as atmospheric CO_2 concentrations and temperature. More in detail, climate change is causing a substantial decrease in rainfall in many regions of the planet such as the Mediterranean area,[1,2] with obvious negative impacts on the availability of water for irrigation. This in turn causes a sharp reduction in the resilience of food systems, thus compromising food security and/or exacerbating the problems of volatility in the prices of agricultural products.

An efficient solution to counter all these problems is to increase the availability of water for agricultural purposes, including the use of treated wastewaters (TWWs) for irrigation. This practice is in fact recommended by international bodies, such as UNESCO,[3] and commonly adopted by many countries that suffer from water scarcity.[4] Furthermore, the growing attention toward the development of a circular approach to the use of environmental resources makes this topic particularly relevant.[5]

However, the reuse of wastewaters (WWs) in agricultural processes poses the risk of potential adverse impacts on nutrients and salinity of soils, as well as biological and chemical concerns.[6,7] Among the adverse effects of this reuse practice, the presence of organic micropollutants (OMPs) in TWWs may negatively affect both the environment and human health, owing to the possible transfer of OMPs from recycled water to soil and crops. In fact, even though wastewater treatment plants (WWTPs) are fundamental elements of modern wastewater management, their treatment chains are often not designed to obtain zero emissions of organic micropollutants[8] and the degradation processes that take place in WWTPs originate a very high number of transformation products, the identity and concentration of which are most of the time unknown.[9]

In order to limit the problem of the transfer of contamination from WW to soil and edible crops, specific regulations have been promulgated in many countries, managing the WW reuse in various fields, including agricultural irrigation. As a recent example, the European Commission has proposed a directive on the minimum requirements for water reuse both in terms of wastewater treatments and concentrations of substances in TWWs, in order to harmonize the national laws of the member states.[10,11] However, OMPs included in the regulations are forcefully a very small number compared to the multitude present in WWTP effluents, and compliance with these regulations does not ensure the absence of risks for the environment and humans.[4] It must be also underlined the lack in regulations concerning the concentration limits for emerging OMPs in fruits and vegetables, being polycyclic aromatic hydrocarbons (PAHs), polychlorinated dibenzo(p) dioxins (PCCDs), polychlorinated dibenzo(p)furans (PCDFs), and dioxin-like polychlorinated biphenyls (DL-PCBs) the only pollutants currently regulated.[12]

Thus, the monitoring of emerging OMPs in TWWs intended for agricultural irrigation, soils irrigated with TWWs, and crops obtained under irrigation with TWWs is of paramount importance for assessing their impact on the environment and human health and providing policy-relevant information.

Many articles dealing with the issue of the transfer of organic contaminants to edible crops due to the TWW reuse in agriculture have been reported in the recent literature. In these studies, the attention is mainly focused on pharmaceutical compounds (PhCs), which have been largely examined in the last 10 years.[13–15] In fact, PhCs are of high concern due to their constant emission into the environment and their high propensity to interact with biological organisms, promoting, for example, in the case of antibiotics, the antimicrobial resistance in bacteria.[16,17] Other investigated OMPs classes include chemicals used for several industrial applications, such as personal care products (e.g., UV-filters and UV-stabilizers) and polymers (e.g., bisphenol A and brominated and perfluorinated compounds).[16,18–20]

As a whole, complications arise in understanding the sorption mechanisms of OMPs on and in soil, the transfer to the crops and the risk assessment for humans.[21,22] As elsewhere reported, these sorption and transfer processes depend on the characteristics of WWs, soils, crops, and physicochemical properties of the compounds considered.[23,24]

This chapter investigates the complex scenario of the transfer of OMPs in soil and crops due to the WW reuse practice through an overview of current literature mainly focusing on studies performed under real agricultural conditions. Specific aims of this overview are (i) obtaining general indications on the environmental fate of the organic compounds studied, (ii) highlighting any knowledge gap both on the compounds and the type of crops examined, and (iii) assessing the risk associated to the ingestion of edible crops produced under irrigation with reclaimed WWs, and (iv) providing possible indications for future research.

The search engines Google Scholar and SciFinder were used for the literature overview concerning the OMPs occurrence in reclaimed WWs and/or soil and/or edible crops, limiting the time interval from 2010 or 2005 to today for PhCs and the other OMPs, respectively. Within the entire dataset reviewed here, the data referring to the OMPs detected both in reclaimed WWs, and in irrigated soils, and in the edible crops grown under WWs irrigation (WW/soil/edible crops systems), have been organized into two tables. Table 1 illustrates the names, abbreviations, CAS numbers, structures and logarithms of the n-octanol/water partition coefficient (logD) of non-regulated OMPs investigated and detected in the whole WW/soil/edible crop systems, whereas Table 2 shows OMPs concentrations found in the systems for various crops. Considering that pH of WWs commonly ranges from 6.5 to 8.5,[5] for each OMP Table 1 shows the most representative structure at pH 7.5, and logD values calculated at pH values of 7 and 8.[25] Note that OMPs mentioned in the text without abbreviation are those that in the articles reviewed here have not been studied in any WW/soil/edible crop system.

2. Use of WWs in agricultural production: Aim and consequences

2.1 Highlights on the use of WWs for irrigation purposes

WW treatment and reuse vary moving from a region to another. In arid and semi-arid regions of developed countries, WWs are treated according to stringent effluent discharge regulations regarding environmental quality, and TWWs is widely used for irrigation, given the increasing demand of water for agriculture and other productive sectors.[26] Since the presence

Table 1 Compound IUPAC name, abbreviation, CAS registry number, chemical structure, pH dependent partition coefficient (logD) and water solubility (g/L) of emerging OMPs analyzed in TWWs and agricultural system.

Compound	Abbreviation	CAS[a]	Chemical structure[b]	logD[c]
Pharmaceuticals				
10,11-epoxycarbamazepine	**(EP-CBZ)**	**36507-30-9**	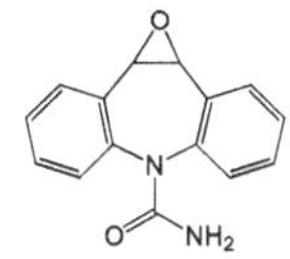	**0.16**
10,11-transdihydroxycarbamazepine	**(DiOH-CBZ)**	**58955-93-4**	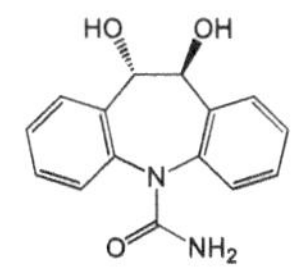	**−0.40**
Atenolol	**(ATE)**	**29122-68-7**	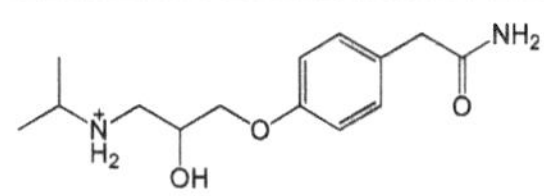	**−2.09**[d] **−1.20**[e]
Caffeine	**(CAF)**	**58-08-2**	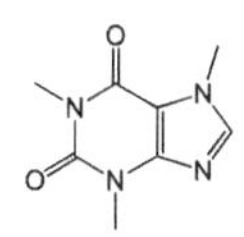	**−0.63**
Carbamazepine	**(CBZ)**	**298-46-4**	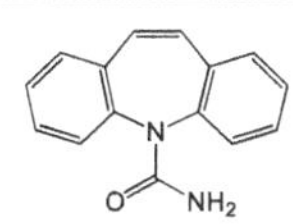	**1.89**
Chloramphenicol	**(CPL)**	**56-75-7**	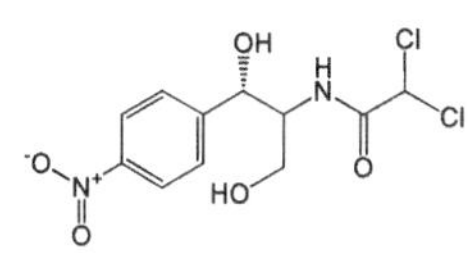	**1.10**
Diclofenac	**(DIC)**	**15307-86-5**	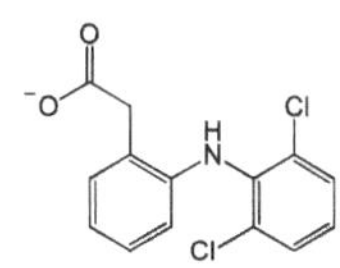	**2.07**[d] **1.18**[e]
Diphenhydramine	**(DPH)**	**58-73-1**	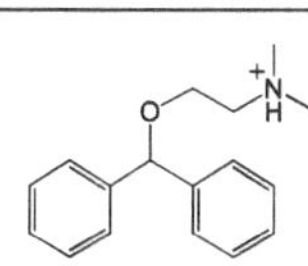	**1.25**[d] **2.17**[e]

Continued

Table 1 Compound IUPAC name, abbreviation, CAS registry number, chemical structure, pH dependent partition coefficient (logD) and water solubility (g/L) of emerging OMPs analyzed in TWWs and agricultural system.—cont'd

Compound	Abbreviation	CAS[a]	Chemical structure[b]	logD[c]
Erythromycin	(ERY)	114-07-8	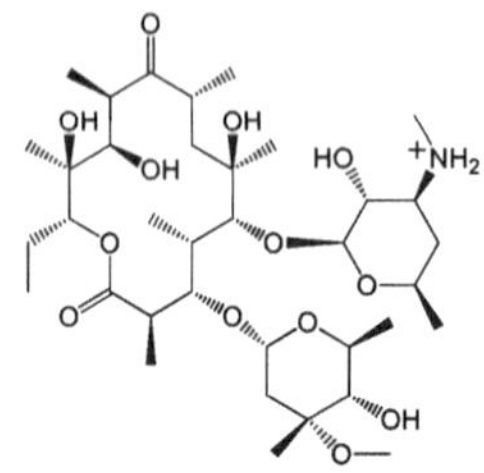	0.81[d] 1.58[e]
Gemfibrozil	(GFZ)	25812-30-0	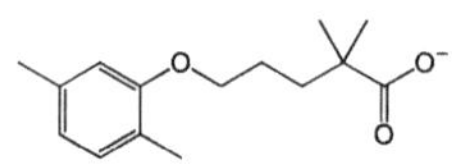	2.07[d] 1.18[e]
Hydrochlorothiazide	(HCT)	58-93-5	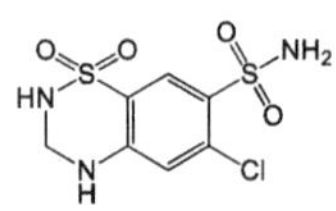	−0.03[d] −0.08[e]
Norfloxacin	(NOR)	70458-96-7	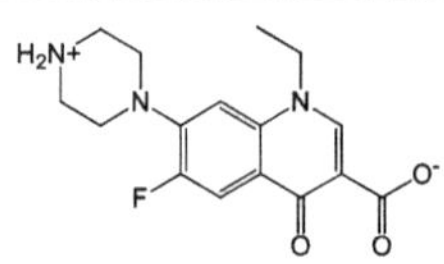	−0.65[d] −0.75[e]
Sulfamethazine	(SMZ)	57-68-1	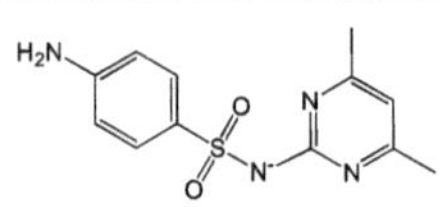	0.15[d] −0.37[e]
Sulfamethoxazole	(SMX)	723-46-6	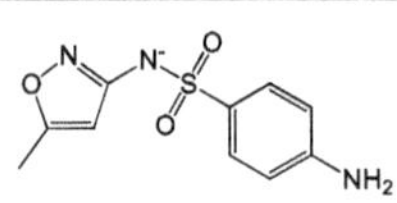	−0.22[d] −0.96[e]
Tetracycline	(TET)	60-54-8	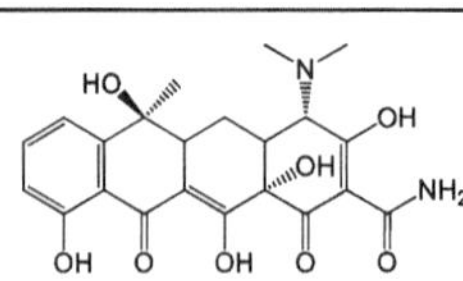	−2.06[d] −2.96[e]
Trimethoprim	(TMP)	738-70-5	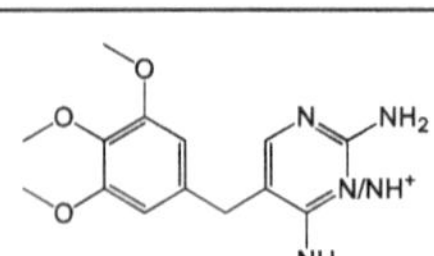	0.27[d] 0.55[e]

Table 1 Compound IUPAC name, abbreviation, CAS registry number, chemical structure, pH dependent partition coefficient (logD) and water solubility (g/L) of emerging OMPs analyzed in TWWs and agricultural system.—cont'd

Compound	Abbreviation	CAS[a]	Chemical structure[b]	logD[c]
Other OMPs				
Ambrettolide	**(ABT)**	**123-69-3**	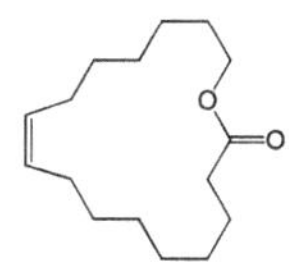	**5.45**
Butylated hydroxyanisole	**(BHA)**	**25013-16-5**	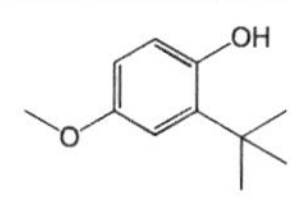	**3.06**
Butylated hydroxytoluene	**(BHT)**	**128-37-0**	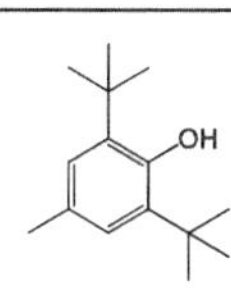	**5.17**
Galaxolide	**(HHCB)**	**1222-05-5**	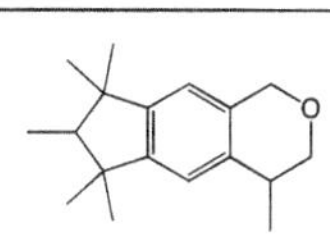	**5.04**
Perfluorobutanoic acid	**(PFBA)**	**375-22-4**	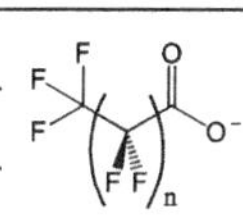	**−0.36**
Perfluoropentanoic acid	**(PFPeA)**	**2706-90-3**		**0.64**
Perfluorohexanoic acid	**(PFHxA)**	**307-24-4**		**1.24[d]** **1.23[e]**
Perfluoroheptanoic acid	**(PFHpA)**	**375-85-9**		**1.97[d]** **1.96[e]**
Perfluorooctanoic acid	**(PFOA)**	**335-67-1**		**2.69**
Perfluorononanoic acid	**(PFNA)**	**375-95-1**		**3.42**
Perfluorobutanesulfonic acid	**(PFBS)**	**375-73-5**	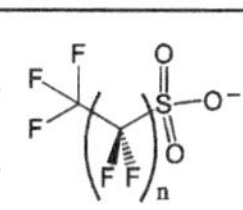	**−0.36**
Perfluorohexanesulfonic acid	**(PFHxS)**	**355-46-4**		**−0.45**
Perfluorooctanesulfonic acid	**(PFOS)**	**1763-23-1**		**1.01**
Phthalates	**–**	**–**	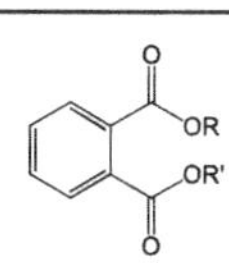	**1.69** **4.69[f]**

Continued

Table 1 Compound IUPAC name, abbreviation, CAS registry number, chemical structure, pH dependent partition coefficient (logD) and water solubility (g/L) of emerging OMPs analyzed in TWWs and agricultural system.—cont'd

Compound	Abbreviation	CAS[a]	Chemical structure[b]	logD[c]
Tonalide	**(AHTN)**	**21145-77-7**	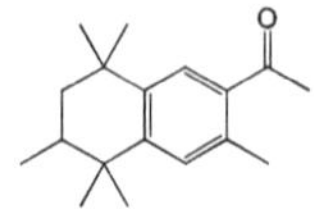	**5.06**
Tributyl phosphate	**(TBP)**	**126-73-8**	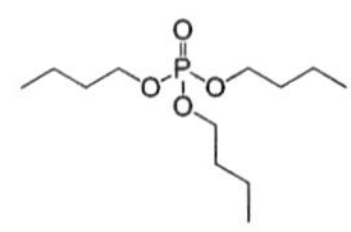	**3.83**
Triclocarban	**(TCC)**	**101-20-2**	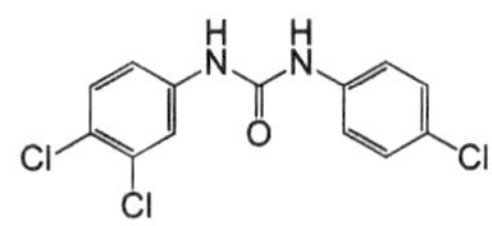	**6.07**
Triclosan	**(TCS)**	**3380-34-5**	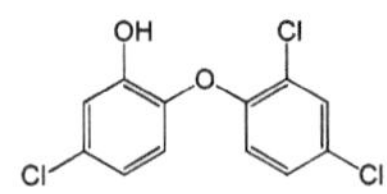	**5.28**[d] **4.93**[e]
Tris(1,3-dichloro-2-propyl) phosphate	**(TDCPP)**	**13674-87-8**	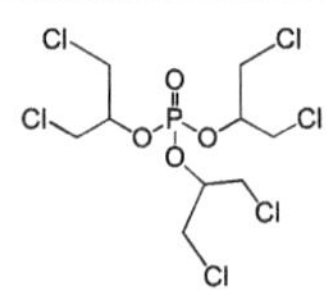	**3.27**
Tris(2-carboxyethyl) phosphine	**(TCEP)**	**5961-85-3**	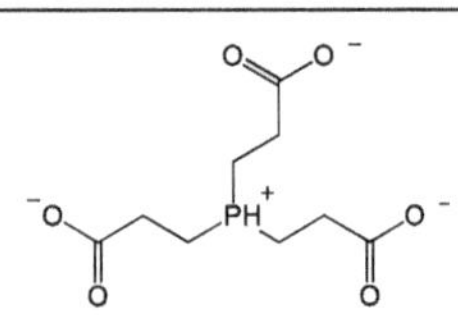	**−5.91**[d] **−5.97**[e]
Tris(2-chloroisopropyl) phosphate	**(TCPP)**	**13674-84-5**	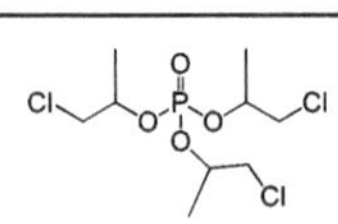	**2.53**

[a]Chemical Abstract Service registry number.
[b]Molecular structure of the most abundant microspecies in common TWWs pH range (6.5–8.5).
[c]Logarithm of the pH dependent octanol–water partition coefficients (logD) has been calculated using Advanced Chemistry Development (ACD/Labs) Software V11.02 (© 1994–2020 ACD/Labs).
[d]logD at pH 7.
[e]logD at pH 8.
[f]Range (minimum–maximum) of logD for the category of phthalates.

Table 2 Occurrence of emerging OMPs in treated wastewaters (TWWs—ng/mL or μg/L), in spiked treated wastewaters (STWWs—ng/mL or μg/L), in agricultural soil (ng/g d.w.) and in the edible portion of the plant (ng/g d.w.).

Compound	TWWs[a]	STWWs[a]	Soil[a]	Plant edible portion[a]	References
Carrot					
EP-CBZ	0.2–0.7		n.a.	7.7[b]	[84]
DiOH-CBZ	2.1–7.2		n.a.	7.4[b]	[84]
CAF	<0.01–1.03	1.5 ± 0.7	8.0–12.5	0.5–4.1	[37]
CBZ	0.5 ± 0.2	1.3 ± 0.5	3–7	6.6–9.1	[37]
	1.2–2.4		n.a.	14[b]	[84]
CPL	2[b]		80[b]	1.62 ± 0.03	[85]
ERY	2[b]		70[b]	1.18 ± 0.02	[85]
NOR	2[b]		90[b]	4.0 ± 0.1	[85]
SMX	<0.01–0.04	0.7 ± 0.4	0.01–0.25	0.04–0.21	[37]
SMZ	2[b]		15[b]	0.3 ± 0.1	[85]
TET	2[b]		145[b]	1.40 ± 0.02	[85]
ABT	0.099–0.497		n.a.	1144[b]	[38]
BHT	<0.192		n.a.	299[b]	[38]
HHCB	0.177–0.451		n.a.	188[b]	[38]
AHTN	0.117–0.226		n.a.	179[b]	[38]
TBP	<0.120		n.a.	85[b]	[38]
Potato					
EP-CBZ	0.2–0.7		n.a.	14	[84]
DiOH-CBZ	2.1–7.2		n.a.	4.7[b]	[84]
CAF	<0.01–1.03	1.5 ± 0.7	1.6–5.6	0.8–1.8	[37]
	0.5–17			30.3	[84]
CBZ	0.5 ± 0.2	1.3 ± 0.5	1.7–6	0.5–0.9	[37]
	1.2–2.4		n.a.	77[b]	[84]

Continued

Table 2 Occurrence of emerging OMPs in treated wastewaters (TWWs—ng/mL or μg/L), in spiked treated wastewaters (STWWs—ng/mL or μg/L), in agricultural soil (ng/g d.w.) and in the edible portion of the plant (ng/g d.w.).—cont'd

Compound	TWWs[a]	STWWs[a]	Soil[a]	Plant edible portion[a]	References
SMX	<0.01–0.04	0.7 ± 0.4	0.01–0.20	0.06–0.20	[37]
Lettuce (leaves)					
ATE	0.23 ± 0.22		1.6[b]	24[b]	[77]
EP-CBZ	0.2–0.7		n.a.	89.4	[84]
	0.03–0.06		n.a.	9–30	[102]
DiOH-CBZ	2.1–7.2		n.a.	41.5[b]	[84]
	0.98–1.84		n.a.	12–37	[102]
CBZ	0.07 ± 0.04		0.3[b]	15[b]	[77]
	0.25–0.64		n.a.	32–89[c]	[102]
	0.025[b]		n.a.	<LOQ–9.09[d]	[86]
	1.2–2.4		n.a.	216	[84]
CPL	2[b]		80[b]	3.20 ± 0.05	[85]
DPH	0.170		n.a.	0.6–1.9[d]	[86]
NOR	2[b]		n.a.	5.7 ± 0.3	[85]
SMX	0.860		n.a.	<LOQ–2.9[d]	[86]
TET	2[b]		145[b]	2.0 ± 0.2	[85]
TMP	0.4[b]		n.a.	<LOQ–1.6[d]	[86]
ABT	0.099–0.497		n.a.	1720[b]	[38]
BHA	<0.160		n.a.	3647[b]	[38]
BHT	<0.192		n.a.	4243[b]	[38]
HHCB	0.177–0.451		n.a.	826[b]	[38]
PFBA	<0.006	4.1	n.a.	5500 ± 1010	[41]
PFPeA	<0.019	4.5	n.a.	3770 ± 853	[41]
PFHxA	0.033[b]	8.7	n.a.	2200 ± 555	[41]
PFHpA	<0.013	8.6	n.a.	873 ± 130	[41]
PFOA	0.009[b]	5.8	n.a.	298 ± 46	[41]

Table 2 Occurrence of emerging OMPs in treated wastewaters (TWWs—ng/mL or μg/L), in spiked treated wastewaters (STWWs—ng/mL or μg/L), in agricultural soil (ng/g d.w.) and in the edible portion of the plant (ng/g d.w.).—cont'd

Compound	TWWs[a]	STWWs[a]	Soil[a]	Plant edible portion[a]	References
PFNA	<0.012	2.7	n.a.	84 ± 17	[41]
PFBS	<0.025	10.9	n.a.	1150 ± 259	[41]
PFHxS	<0.002	4.0	n.a.	188 ± 51	[41]
PFOS	0.003[b]	1.4	n.a.	57 ± 20	[41]
TBP	<0.120		n.a.	4312[b]	[38]
TCS	0.021 ± 0.009		3.6[b]	6.9[b]	[77]
TDCPP	0.491[b]		n.a.	110.9–195.9[d]	[86]
TCEP	0.287[b]		n.a.	13.6–24.2[d]	[86]
TCPP	0.711[b]		n.a.	27.1–52.6[d]	[86]
Radish (leaves)					
ATE	0.23 ± 0.22		0.6[b]	2[b]	[77]
CBZ	0.07 ± 0.04		0.3[b]	42[b]	[77]
CPL	0.04–0.07		8–22	8–18	[81]
NOR	0.015–0.025		20–66	5.7–21	[81]
SMZ	0.005–0.015		1.3–4	<0.8–1.9	[81]
TET	0.07–0.12		<4.4–22	<4.4–9.2	[81]
TCS	0.021 ± 0.009		0.3[b]	14.8[b]	[77]
Rocket (leaves)					
EP-CBZ	0.2–0.7		n.a.	33.5[b]	[84]
CBZ	1.2–2.4		n.a.	61[b]	[84]
Parsley (leaves)					
EP-CBZ	0.2–0.7		n.a.	21[b]	[84]
DiOH-CBZ	2.1–7.2		n.a.	11.0[b]	[84]
CBZ	1.2–2.4		n.a.	91[b]	[84]

Continued

Table 2 Occurrence of emerging OMPs in treated wastewaters (TWWs—ng/mL or μg/L), in spiked treated wastewaters (STWWs—ng/mL or μg/L), in agricultural soil (ng/g d.w.) and in the edible portion of the plant (ng/g d.w.).—cont'd

Compound	TWWs[a]	STWWs[a]	Soil[a]	Plant edible portion[a]	References
Cabbage (leaves)					
EP-CBZ	0.2–0.7		n.a.	19[b]	[84]
CAF	1.4[b]		96[b]	12[b]	[44]
CBZ	1.2–2.4		n.a.	79[b]	[84]
CPL	0.04–0.07		8–22	2.6–13	[81]
GFZ	0.255[b]		58[b]	75[b]	[44]
NOR	0.015–0.025		20–66	9.6–24	[81]
TET	0.07–0.12		<4.4–22	4.0–10.1	[81]
Spinach (leaves)					
CPL	0.04–0.07		8–22	<2.4–10.1	[81]
NOR	0.015–0.025		20–66	8.6–22	[81]
TET	0.07–0.12		<4.4–22	<4.4–6.3	[81]
Tomato (fruit)					
EP-CBZ	0.1[b]		0.1[b]	1.0[b]	[80]
DiOH-CBZ	1.07–1.47		n.a.	1.9–5.9	[102]
CAF	0.7[b]		2.5[b]	6.0–20	[80]
CBZ	1.2–2.4		n.a.	5.0[b]	[84]
	0.28–0.63		n.a.	1.0–1.6	[102]
	0.2[b]		0.4[b]	1.2–1.7	[80]
	1.1 ± 0.1		0.2–2.3	0.9–1.2	[78]
CPL	2[b]		80[b]	4.3 ± 0.1	[85]
DIC	0.026–0.049[d]		0.06–0.35[d]	1.28–11.6[c]	[21]
HCT	2.758h		0.2[b]	2.0–4.0	[80]
NOR	2[b]		90[b]	14.0 ± 0.3	[85]
SMX	0.025–0.055[c]		0.38–0.98[d]	0.26–5.25[c]	[21]

Table 2 Occurrence of emerging OMPs in treated wastewaters (TWWs—ng/mL or μg/L), in spiked treated wastewaters (STWWs—ng/mL or μg/L), in agricultural soil (ng/g d.w.) and in the edible portion of the plant (ng/g d.w.).—cont'd

Compound	TWWs[a]	STWWs[a]	Soil[a]	Plant edible portion[a]	References
TET	2[b]		145[b]	2.6 ± 0.4	[85]
TMP	0.022–0.073[c]		0.15–0.62[c]	0.11–3.40[c]	[21]
Cucumber (fruit)					
CAF	0.6 ± 0.3		1.9–3.0	2.2–5.4	[78]
CBZ	1.1 ± 0.1		0.5–4.3	7.6–13.4	[78]
Zucchini (fruit)					
EP-CBZ	0.2–0.7		n.a.	7.1[b]	[84]
CBZ	1.2–2.4		n.a.	6.8[b]	[84]
Eggplant (fruit)					
EP-CBZ	0.2–0.7		n.a.	12[b]	[84]
DiOH-CBZ	2.1–7.2		n.a.	14[b]	[84]
CBZ	1.2–2.4		n.a.	32[b]	[84]
Strawberry (fruit)					
DPH	0.170[b]		n.a.	1.0–1.5	[86]
SMX	0.860[b]		n.a.	8.2–17.5	[86]
TMP	0.400[b]		n.a.	17–39	[86]
PFBA	<0.006	4.1	n.a.	2650 ± 48	[41]
PFPeA	<0.019	4.5	n.a.	1070 ± 81	[41]
PFHxA	0.033[b]	8.7	n.a.	258 ± 21	[41]
PFBS	0.025[b]	10.9	n.a.	18 ± 6	[41]
TCC	0.01[b]		n.a.	<LOQ–0.09	[86]
TDCPP	0.4[b]		n.a.	136.3–161.7	[86]
TCPP	2[b]		n.a.	20.7–29.7	[86]
Pepper (fruit)					
CBZ	1.2–2.4		n.a.	8.3[b]	[84]

Continued

Table 2 Occurrence of emerging OMPs in treated wastewaters (TWWs—ng/mL or μg/L), in spiked treated wastewaters (STWWs—ng/mL or μg/L), in agricultural soil (ng/g d.w.) and in the edible portion of the plant (ng/g d.w.).—cont'd

Compound	TWWs[a]	STWWs[a]	Soil[a]	Plant edible portion[a]	References
Rice (fruit)					
CPL	0.04–0.07		8–22	6.4–20.4	[81]
NOR	0.010–0.075		24–67	10.4–13.8	[81]
TET	0.07–0.12		<4.4–22	5.6–8.5	[81]
Maize (fruit)					
ATE	0.23 ± 0.22		1[b]	0.8[b]	[77]
CBZ	0.07 ± 0.04		0.8[b]	3.8	[77]
CPL	0.04–0.07		8–22	15–22	[81]
NOR	0.010–0.075		24–67	10–17	[81]
TET	0.07–0.12		<4.4–22	<1.1–6.6	[81]
TCS	0.021 ± 0.009		0.3[b]	4.2[b]	[77]

[a]Values are reported as follow: mean ± standard deviation or ranges.
[b]Measured concentration value (no standard deviation available).
[c]Range of mean values on 3 years experiment with two different TWWs.
[d]Range of mean values calculated for soils having 0.4%, 2% and 6% organic carbon content of soil.
n.a., not available; n.d., not detected.

of OMPs in WWTP effluents depends on the treatments adopted, the European Commission established preventive measures to limit the environmental and human health impact due to TWW reuse for edible crop irrigation.[4] More in detail, four classes of TWWs have been foreseen for the reuse in the agricultural sector, all of them handled at least with secondary treatment and disinfection, and then assigned to specific categories of crops, depending on the expected impact of TWWs on the agricultural products.[11]

In general, when secondary treatment is applied in WWTPs, biodegradation promotes the removal of OMPs up to 90%.[27] Unfortunately, it is worth noting that in many cases negative balances are observed, indicating the formation of parent compounds through de-conjugation, or accumulation of the substances during treatment.[28] Depending on the influent WWs, WWTPs can adopt tertiary treatments based on advanced oxidation (e.g., ozonation) and/or adsorption (e.g., activated carbon) technologies, capable to promote a further reduction of recalcitrant OMPs compared to secondary stages.[29,30]

In any case, a "zero emission" of OMPs is an unrealistic depuration goal. Such an objective appears even farther from being achieved in the rural areas of developing countries, where WW treatment is limited and untreated WWs are also used for irrigation by small farmers, who cannot provide efficient tools to optimize WW quality.[26] Of course, when crops are irrigated with untreated WWs, the chemical risk associated to the presence of OMPs is expected to increase.[31]

Based on the aforementioned considerations, the accurate investigation of OMPs in WW intended to be reused for irrigation of edible crops is of paramount importance. In this regard, the overview of literature data evidences a strong tendency to the targeted investigation of those chemicals, which are expected to be present in WWs,[32] whereas the attempt to perform a comprehensive overview of selected OMPs together with their transformation products is almost completely missing.

2.2 OMPs in WWs used for agricultural irrigation

A wide group of selected OMPs has been investigated in WWs reused for irrigation, giving special attention to those chemicals that, due to their diffusion, persistence, and toxicity pose major environmental and health risks.[28,33]

More in detail, examining the literature regarding the presence of OMPs in WW/soil/edible crop systems (Table 2), PhCs have provoked increasing concern, due to their distribution, lack of precise regulations and negative effects on the environment.[17] As a general perspective, depending on the area investigated and season,[34] PhCs (e.g., antibiotics, nonsteroidal anti-inflammatory drugs, hormones etc.) concentrations in WWs are mostly in the range of 10^{-1}–10^{2} μg/L.[14] As regards antibiotics, the classes most commonly investigated in the WW/soil/crop systems are fluoroquinolones, sulfamides and macrolides that in conventional activated sludge WWTPs show highly variable removals (8–73%),[14] probably due to different behaviors of the microbial communities colonizing the biological oxidation tanks. On the other hand, where tertiary treatments are applied after biological oxidation, their removal efficiency is brought close to 100%.[35,36] Antibiotic concentrations in WWs reused for irrigation vary depending on the origin of the sewage and the type of treatment adopted.[14] CPL, CAF, TET, TCS and CBZ have been also investigated very frequently in WWs, due to their broad diffusion and recalcitrant behavior during secondary treatment.[37–39]

Emerging OMPs other than PhCs often investigated in WWs for agricultural reuse, belong to the groups of personal care products (PCPs) (e.g., fragrances and antioxidants) and additives employed in various industrial products. Such chemicals are widespread monitored due to their recalcitrance and potential endocrine disruptive activity, and occur in WWTP effluents generally in the ng/L to low μg/L range.[20] In detail, two of the most frequent occurring categories are synthetic musk fragrances and perfluoroalkyl acids (PFAAs), both of them highly recalcitrant to activated sludge treatments. These OMPs occur in WWTP effluents at very different concentration ranges, depending on WW origin and treatment, i.e., 0.1–4 μg/L [40–42] and 0.001–4 μg/L,[43,44] for fragrances and PFAAs, respectively. In addition, PFAAs have often shown an increasing concentration trend during treatment and only advanced oxidation and activated carbon treatments may be effective for their removal.[44,45]

Moreover in WWTPs, OMPs transformation products occur as a result of abiotic and biotic degradation processes,[46] but as mentioned previously they are seldom investigated. An interesting example, which will be discussed with some details in Section 4.3, is the occurrence of CBZ transformation products, being them by far the predominant metabolites detected in TWWs, due to excretion and biological oxidation in WWTPs.[47]

3. Behavior of OMPs in soil

3.1 Highlights on OMPs occurrence in soil

OMPs transfer to soil can mainly occur through irrigation with TWWs or application of biosolids as amending agents.[21,48] However, the use of TWWs in agriculture is commonly considered the main source of soil contamination by OMPs. In fact, the soil must be irrigated continuously for the entire duration of the crop vegetative period until harvest, while the sludge is applied to the soil only periodically. Moreover, OMPs in TWWs are in soluble form, being therefore characterized by a greater mobility in soil and a consequently higher bioavailability.[13] Conversely, sludge acts as a source but also as a sink of OMPs, due to its organic matter content.[49–51]

OMPs present in TWWs, once conveyed on the soil, are subject to very complex sorption/desorption and transformation processes, depending on their physicochemical properties and characteristics of the soil, as well as of TWWs used for irrigation.[21] In this latter regard, the concentration of dissolved organic matter (DOM), mainly consisting in proteins, carboxylic and polycarboxylic acids, polysaccharides, and lipids,[52] play an important role on the fate of OMPs in soil. More in detail, DOM may strongly interact with OMPs and compete with soil adsorption sites, thus increasing OMPs solubility and mobility in soil. However, a decrease of OMPs mobility in soil has been also reported, because of the strong interactions between DOM-OMPs complexes and soil.[13,53] The sorption/desorption processes of OMPs in soil are governed by the specific interactions occurring among functional groups of OMPs, DOM in TWWs, and organic matter in soil (SOM) and are therefore strongly dependent on the peculiar characteristics of contaminants, TWWs and soil.[54]

3.2 Influence of soil properties on OMPs sorption/desorption and degradation processes

The effect of DOM on the sorption/desorption processes during irrigation has been extensively studied. For example, the sorption of CBZ (logD = 1.89), which is in neutral form regardless the TWWs pH value, was not influenced by the presence of DOM on a clay soil with SOM = 1.04%. On the other hand, the preventive sorption of DOM on the same soil, which alters its content in OC, leads to an increase of CBZ retention of about 18–40%, depending on the compound concentration.[55] In the case of sulfapyridine (logD = 0.4, $pKa_1 = 2.1$, $pKa_2 = 8.5$), which occurs in neutral/anionic form at pH 6.5–8.5, the sorption on low-SOM clay and sandy-loam soils (<0.1%) seems to be mainly linked to chemisorptive polar interactions and hydrogen bonds between functional groups of the molecule and soil mineral surface, as elsewhere reported.[56,57] Furthermore, when DOM is pre-adsorbed on soil, the sulfapyridine desorption is facilitated probably due to a concentration- and/or composition-dependent competitive binding.[57] However, the enrichment in SOM by DOM after 10 years irrigation with TWWs has not been highlighted in clay and sandy clay soils, thus making doubtful the impact of TWWs on the amount of SOM.[58]

Similarly to what has been reported for DOM, the sorption/desorption processes of OMPs may be influenced by SOM. In detail, the mobility of OMPs characterized by a wide range of hydrophobicity decreases with increasing the content of SOM, due to the sorption phenomena. This is the case of hydrophobic and predominantly neutral aromatic compounds such as bisphenol A (logD = 3.6, pKa = 10) and TCS (logD = 4.9–5.3, pKa = 7.7),[59] but also of more hydrophilic compounds such as CBZ and ATE, which contains aromatic rings and peculiar polar moieties that can strongly interact with polar and/or aromatic groups occurring in SOM.[60]

A decreasing mobility with increasing SOM can be highlighted also for PFAAs, which are anionic molecules characterized by a strong variability in hydrophobicity as a function of chain length and type of acidic group.[43]

A more complex picture is reported for sulfonamide antibiotics in relation to SOM. In fact, these chemicals undergo to protonation/deprotonation of the sulfonamide nitrogen, right in the pH range commonly found for TWWs. Accordingly, different sorption abilities have been observed for sulfathiazole (logD = −0.1, pKa = 7.2) even within relatively low pH variations in loam, sandy-loam and loamy-sand soils.[61] However, a preferential partition of sulfonamides in soil, as well as the increase of K_D with increasing SOM, have been reported at pH values corresponding to the significant occurrence of the anionic species. This finding may be ascribed to the formation of complexes with metal cations in solution.[62] Interestingly, neutral molecules containing polar functional groups, such as CPL (logD = 1.1, most acidic pKa = 11), CAF (logD = −0.6, most basic pKa = 0.5), tinidazole (logD = −0.3, most basic pKa = 2.3), and metronidazole (logD = −0.14, most basic pKa = 2.6) are capable to interact with the polar and/or aromatic groups occurring in SOM, being consequently retained onto soil despite their low hydrophobicity.[63] It should also be noted that, SOM, being constituted of negatively charged humic substances,[64] is also capable to exert a repulsive effect against OMPs that have a dissociated carboxylic moiety at around neutral pH. Among them, acidic antibiotics (e.g., SMX, logD = −0.2, pKa = 5.8) and nonsteroidal anti-inflammatory drugs (e.g., naproxen, logD = 0.7, pKa = 4.8), occur in dissociated anionic form in TWWs used for irrigation,[21,65] thus increasing their mobility in soils.[66,67]

Even though SOM plays a fundamental role in the complex mechanisms regulating the sorption of OMPs in soils irrigated with TWWs, soil textural properties may strongly affect OMPs mobility.[21] In particular, clay percentage is of crucial importance, since its occurrence determines the number of cationic exchange sites in soil, which promote the sorption of positively charged molecules such as tetracycline (logD = −2.1; pKa_1 = 4.5; pKa_2 = 11.2) or doxycycline (logD = −0.9; pKa_1 = 4.5; pKa_2 = 10.8) that otherwise will permeate soil layers, due to their high polarity.[68] Physical characteristics of soil such as surface area of soil particles and macropores volume of bulk soil also affect the interactions between OMPs and soil itself.[66,68,69] In this regard, the use of TWWs for soil irrigation results in a wide range of negative effects on soil structure and composition, which may involve the decrease in soil permeability (i.e., "clogging effect"), as well as the increase in salinity and electrical conductivity (EC). The decrease of soil permeability results in a lower infiltration capacity of OMPs, with a consequently lower contamination of the deeper soil layers and of groundwater, but at the same time the potential increase of OMPs transfer to crops. Moreover, soil salinization commonly involves the decrease of crop yield.[70] The extent of these effects depends on native properties of soils and TWW quality, the latter in turn related to the type of wastewater treatment.[71,72]

Sorption processes of OMPs in soil may also affect their degradation, being the properties of OMPs and the amount and characteristics of DOM and SOM relevant parameters in this context. As a general consideration, the increase in SOM should increase the half-life of OMPs due to the increase of OMPs sorption in soil. In fact, biodegradation should occur more rapidly in soil solution than in bulk soil. Accordingly, OMPs with high to medium polarity such as naproxen (logD = 0.7), ibuprofen (logD = 0.9), CBZ (logD = 1.8), and DIC (logD = 2) showed a significant reduction of the degradation rate with increasing SOM. Conversely, more hydrophobic compounds (e.g., TCS), which probably are highly retained even in soil with low SOM content, show very similar half-lives irrespective of the SOM value.[59,73] However, in some cases, mixed information are reported in literature. For example, Xu and co-workers[59] reported an increase of bisphenol A (logD = 3.6) half-life from 1 to 5 days by increasing SOM from 0.6% to 5.5%, whereas Yu et al.[73] found very similar degradation rates in soils with 0.4–3% of SOM.

4. Transfer of organic micropollutants in edible crops

4.1 Mechanisms of plant uptake and translocation of OMPs

The transfer of OMPs from TWWs to edible crops has been widely reported in the literature,[13,74–77] although most studies are performed in experimental conditions other than real ones (e.g., in hydroponics or irrigating the soil with wastewater spiked with OMPs at concentrations significantly higher than those actually found in TWWs) and only few works take into consideration the whole TWWs/soil/edible crop system.[37,39,46,78–83] In these latter studies, the evidence of the transfer of several OMPs in different crops is highlighted (see Table 2). The transfer mechanisms are linked both to the different factors mentioned in Section 3.1 (i.e., characteristics of the

OMP, TWW, and soil that determine the complex equilibria within the soil matrix) and to the multifaceted mechanisms that regulate the uptake and translocation of OMPs in plants.

Even though the mechanisms regulating the fate of OMPs in plants are not fully clarified,[75] some general findings have been highlighted in this regard.[74–76] The absorption of OMPs by plants and their translocation to the various plant organs through the vascular system (i.e., phloem and xylem) requires that the contaminants penetrate the various levels of the root system (i.e., epidermis, cortex, endodermis, and casparian strip).[74] Within this complex absorption mechanism, the presence of a charge on the molecule plays a fundamental role. In fact, anionic OMPs should undergo repulsion effects due to the negative electrical potential at the cell membrane on plant cells. Therefore, anionic OMPs should be less absorbed than cationic ones, being the latter affected by the charge attraction of biomembranes and/or other ionic interaction mechanisms. Moreover, neutral molecules seem to be more rapidly absorbed and more easily translocated than ionic ones by phloem and xylem.[13,74] Indeed, an increasing transfer of OMPs in the order anionic < cationic < neutral was clearly demonstrated for tomato and cucumber grown in soils with different textures and SOM contents, spiked with several PhCs.[80]

The presence of a charge in the molecule—and more in general its polarity—influences the fate of OMPs also inside the plant. More in detail, the most hydrophobic compounds tend to accumulate in the lipid part and to be retained by the roots, while the more hydrophilic ones have greater mobility toward the aerial part through the xylem.[75]

The characteristics of the soil solution and the concentrations of OMPs contained therein are also a relevant aspect for the uptake of contaminants. In this regard, the fraction of OMPs contained in the interstitial water has proven to be the most available for the plant and with the greatest correlation with the absorbed fraction.[84] Nevertheless, most researches focus on OMPs occurrence in the soil bulk, without investigating specifically the soil solution.

Moreover, the peculiar plant genetic characteristics result in profound metabolic differences (e.g., plant growth and transpiration rates, proteins found in the xylem sap, and differences in tissue lipid contents), which in turn affect plant uptake and translocation of OMPs.[74,76] OMPs may also undergo degradation inside plant tissues, thus making still more complex the understanding of the fate of OMPs in plants.[85]

4.2 Occurrence of OMPs in edible crops

As illustrated in Table 2, the occurrence of OMPs has been evidenced in various edible crops. The detected OMPs cover a wide range of physicochemical characteristics (see Table 1), belonging to the most diverse product categories.

PhCs have been by far the OMPs most studied in plants. More in detail, among the several PhCs investigated, CBZ has been found in most cases at higher concentrations compared to the other detected drugs, in all the edible parts of crops (i.e., in root or leaves or fruits), in agreement with its physicochemical properties (see Table 1). In fact, the lack of charge and the relative polarity of the molecule make it easily assimilated by the roots and translocated in the aerial part of plants.[79,86] Interestingly, CAF, which is a neutral molecule like CBZ, even though much more polar ($\log D = -0.63$), has been detected at twofold lower concentration values in root vegetables (e.g., potato and carrot)[39] compared to leafy vegetables (e.g., cabbage) and fruits (e.g., cucumber and tomato).[46,80,82,86] The data shown in Table 2 evidence an interesting trend for antibiotic occurrence in the edible portion of plants. In fact, regardless the type of crop and in the presence of similar WWs concentrations, the occurrence in the edible portion of the following antibiotics were in the order NOR > CPL ≥ TET > ERY.[83,87] More in detail, ERY was often undetected in crop samples probably due to its high molecular weight (733.9 Da) and the presence of a net charge on the molecule, in comparison with the other antibiotics, which are overall uncharged and have a lower molecular weight. Interestingly, SMX has been also detected in various edible crops (i.e., carrot, potato, lettuce, tomato, and strawberry),[37,39,88] notwithstanding it carries a negative charge and should therefore undergo repulsion effects at the root level, due to the negative electrical potential at the cell membrane on plant cells. The transfer of SMX to plant organs is probably promoted by electrostatic repulsion forces between the soil matrix and the molecule, as suggested by the higher absorption observed in root vegetables (i.e., carrot and potato) grown in a sandy-loam soil with 1.3% of SOM, compared to other soils with a much lower percentage of organic matter.[39]

In agreement with the uptake and translocation mechanisms discussed in Section 4.1, hydrophobic neutral compounds ($\log D > 4$) such as synthetic musks (e.g., ABT, HHCB and AHTN), and antioxidants (e.g., BHT and BHA), have been found in carrots and lettuce at high concentration levels (10^2–10^3 ng/g d.w.), despite the short period of irrigation.[40] Anionic molecules, such as PFAAs also showed a high accumulation (up to 1000 ng/g d.w.)

in leafy vegetables and fruits with high water content (i.e., lettuce and strawberry) when high concentrations (μg/L) of perfluorinated compounds were spiked in TWWs used for their irrigation.[43] The accumulation of each PFAA was linearly correlated with TWW concentration and was explained by passive transport as possible transfer mechanism through plants, with the preferential uptake of short-chain PFAAs. However, the accumulation of these compounds in fruits (i.e., strawberry and olive) irrigated with unfortified WWs (PFAAS concentrations in the ng/L order of magnitude), is generally lower than 1 ng/g d.w. and in some cases even lower than PFAAs concentrations in commercially available fruits.[89]

Organophosphorus flame retardants (see Table 2), which are uncharged compounds, also appear to be easily incorporated into plant tissues. This finding has been confirmed by evaluating the uptake of these chemicals in lettuce and strawberries.[40,88] More in detail, the uptake of organophosphorus chemicals seems to be driven mainly by their hydrophobicity (see Table 1), since the following trend has been evidenced: TBP $\gg$ TDCPP $\geq$ TCPP $>$ TCEP. The concentration values found in lettuce were slightly higher compared to those reported for strawberry fruits, suggesting that the transfer of these chemicals is limited by passive diffusion, thus accumulating predominantly in leafy plant tissue.[88] Similar results were obtained in the investigation of triclosan (logD $\sim$ 5) transfer, for which the uptake of the uncharged molecule is preferential in leaves rather than fruits. Anyway, the low triclosan concentrations in leaves, compared to what has been found in roots, suggested that such hydrophobic chemical strongly binds to roots surfaces hindering the chemical's transport along the plant body.[79]

4.3 Assessment of the risk associated with the intake of food contaminated with OMPs

The risk assessment is a systematic process for describing and quantifying the risk associated with hazardous substances, processes, actions, or event.[90] The agricultural food supply chain that involves the use of TWWs for irrigation purposes undoubtedly represents an important route of human exposure to contaminants and therefore a risk to human health.[91] In this work, the risk evaluation is applied to the estimation of potential hazardous effects due to the ingestion of food crops produced under irrigation with TWWs and contaminated with OMPs. In particular, according to the World Health Organisation[92] the dietary exposure (D.E.) to a certain contaminant in a certain food has been calculated based on the equation reported below (Eq. 1).

$$D.E. = C \times F.I. \quad (1)$$

In the Eq. (1), C is the concentration of the contaminant in the fresh weight (f.w.) food (ng/g f.w.), while F.I. is the f.w. mass of food ingested daily per kg of body weight (B.W.), expressed as g f.w./kg B.W.·d. Hence, D.E. has unit ng/kg B.W.·d. When available on a f.w. basis, C values were taken from data in the original manuscripts. Conversely, when C values were expressed on a d.w. basis, the conversion to f.w. concentrations was performed by using moisture percentages reported in an USEPA database.[93] F.I. data regarding the estimated ingestion by the whole population of all fruits and vegetables considered here, were taken from the same aforementioned database.[93] For zucchini and rocket, which were nor mentioned in the USEPA database, F.I. values were obtained from Aryal and Reinhold[94] and Duri et al.,[95] respectively, whereas the moisture contents of both vegetables were taken from the CREA database of food composition.[96]

The resulting dietary intakes are then compared with health-based guidance values available in literature for food contaminants of concern, which account for the maximum amount of a chemical that can be ingested on a daily basis without any significant increase of the risk for the human health.[92] Examples of reference values include health-based guidance values for setting safe levels of chronic exposure in humans such as (i) acceptable daily intake (ADI) for food and feed additives, pesticides and food contact materials, (ii) tolerable upper intake levels (UL) for vitamins and minerals, and (iii) tolerable daily intake (TDI) for contaminants.[97] These reference values are calculated based on toxicological studies on animal models, applying a safe factor to the calculation, which refers to the uncertainty in the data extrapolation among different living species.[92]

For chemical compounds characterized by unknown toxicity, the principle of the threshold of toxicological concern (TTC) is applied.[98–100] The TTC approach is a risk assessment based on the idea that a *de minimis* threshold risk of exposition to human health exists, which includes a broad group of chemical substances and structures with a low level of occurrence in foodstuff such as flavoring substances, food additives, pesticides, PCPs and PhCs.[100–102] All these compounds are grouped in categories of increasing concern for human health, named Class I, II, and III according to the Cramer classification[103] and in potential DNA-reactive mutagens[101] and their TTC can be found using the Toxtree software, a database facilitating the application of the Cramer scheme.[101]

The TTC risk assessment approach has been widely adopted for estimating the human health exposure related to the consumption of fruits and vegetables containing emerging OMPs due to the irrigation with TWWs.[37,39,82,86]

As previously mentioned, neutral and polar molecules exhibit a greater mobility toward the plant aerial part, while anionic compounds show a lower translocation since they are subject to repulsion effects due to the negative electrical potential at the cell membrane on plant cells. Among the large group of OMPs investigated for their occurrence in fruits and vegetables irrigated with TWWs, CBZ and its metabolites EP-CBZ and DiOH-CBZ represent interesting examples of non-ionic, polar molecules (see Table 1), which have been detected in many edible crops (e.g., strawberry, lettuce, and carrots) (Table 2). As a further relevant example of OMPs widely determined in fruits and vegetables, sulfamethoxazole (SMX) is an anionic antibiotic molecule characterized by high hydrophilicity (see Table 1). Accordingly, these molecules can be considered as models to understand the impact of OMPs transfer in edible crops due to TWWs reuse for their irrigation. In this regard, CBZ and DiOH-CBZ belong to Class III, which include substances with "chemical structures that permit no strong initial presumption of safety or may even suggest significant toxicity or have reactive functional groups."[101] For all the Class III substances, TTC is taken equal to 1500 ng/kg B.W.·d. Conversely, EP-CBZ and SMX, being included by EFSA in the list of substances considered as "potential DNA-reactive mutagens and/or carcinogens," has a TTC of 2.5 ng/kg B.W.·d.

Fig. 1 illustrates the D.E. values for CBZ (Fig. 1A), DiOH-CBZ (Fig. 1B), and EP-CBZ (Fig. 1C) in various TWWs-irrigated edible crops, compared to TTC thresholds, whereas Fig. 2 shows the same information for SMX. D.E. values for CBZ, DiOH-CBZ, and SMX were in all cases very low compared to TTC reference values (about two-three magnitude orders lower), suggesting that the exposure to these OMPs through the consumption of crops irrigated by TWWs is not critical for human health. These considerations apply to most data presented in Table 2, thus suggesting a negligible risk due to the ingestion of each individual OMP present in each crop. However, as illustrated by Fig. 1C, a different picture is observed for EP-CBZ, which exhibits D.E. of the same magnitude order of TTC (2.5 ng/kg B.W.·d) for selected samples of potato (1.58 ng/kg B.W.·d) and parsley (1.37 ng/kg B.W.·d).[86] Moreover, as a general consideration, it must be remarked that this risk estimation is based for each compound on the assessment of ingestion through one food at a time, while obviously

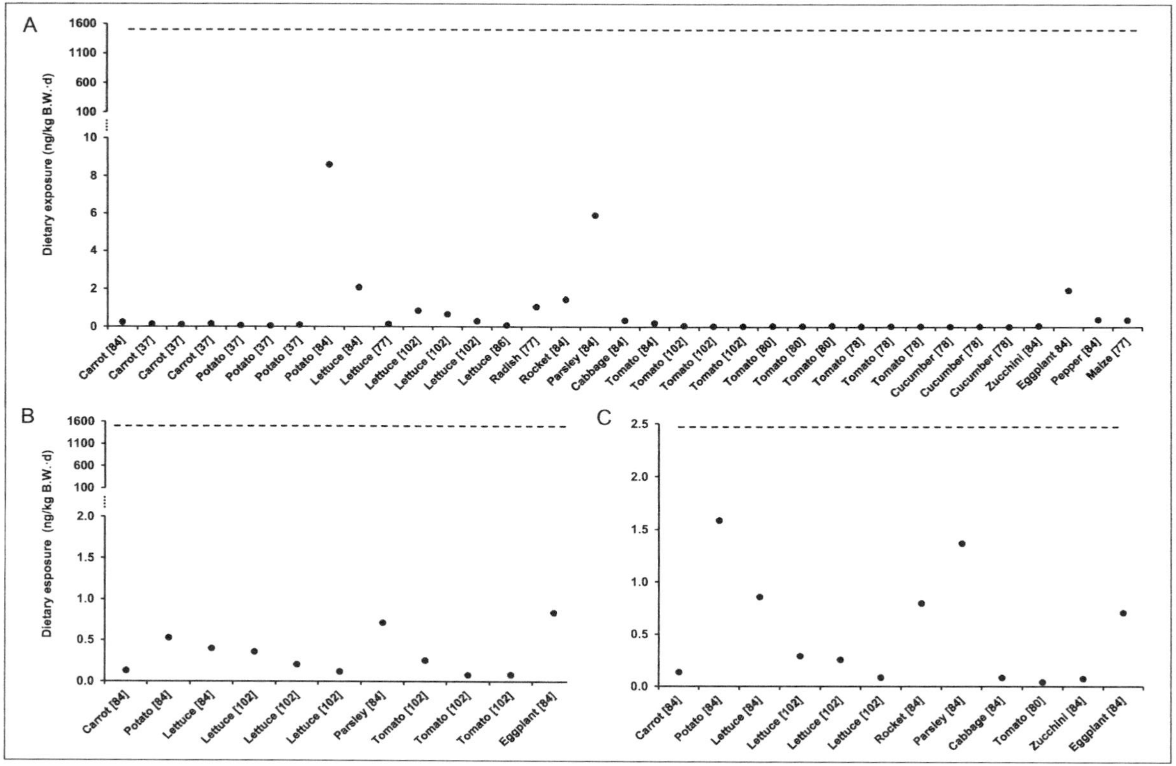

Fig. 1 Dietary exposures (ng/kg B.W.·d) of (A) CBZ, (B) DiOH-CBZ and (C) EP-CBZ. Numbers in square brackets are the bibliographic references from which data have been collected for the evaluation of the dietary exposures. The dashed line represent the TTC reference value.

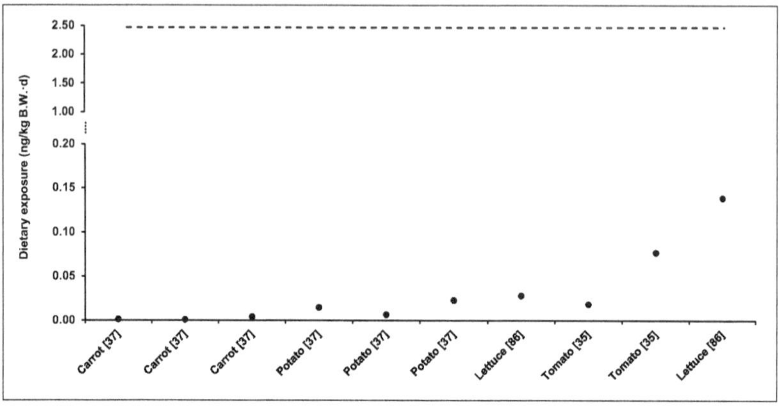

Fig. 2 Dietary exposures (ng/kg B.W.·d) of SMX. Numbers in square brackets are the bibliographic references from which data have been collected for the evaluation of the dietary exposures. The dashed line represent the TTC reference value.

it can happen to ingest multiple foods contaminated by the same OMP. In addition, this calculation does not take into account the cumulative risk deriving from the ingestion of multiple compounds present in the same food or in different foods. Therefore, in our opinion, the existence of a risk for human health arising from the intake of agricultural products irrigated with TWWs cannot be excluded and should be conversely given due consideration.

5. Concluding remarks

TWWs is already and will probably be an even greater resource for the production of edible crops in the near future. It is therefore of great importance that TWWs are produced using treatment systems with efficient and reproducible performances. In this context, the adoption of tertiary treatment systems should be widely promoted, so as to significantly reducing concentrations of OMPs, compared to what is observed in the effluent of biological secondary treatments.

Although modern chromatographic techniques hyphenated with high-resolution tandem mass spectrometry allow untargeted investigations suitable for identifying the degradation products of OMPs, the panorama of data currently available in TWWs, in soil and in crops is almost completely limited to parent compounds. Conversely, the analysis of degradation products is conducted only rarely (with the exception of CBZ), thus representing a gap of knowledge to be filled.

It is important to remember that the exposure to OMPs due to their ingestion depends not only on its concentration in food, but also on the extent of F.I., which is highly variable depending on the fruit or vegetable considered. Therefore, it could be planned the use of TWWs only for the irrigation of edible crops which have a low to medium daily consumption and are not used as a basis for the formulation of largely consumed foods, such as wheat. In this regard, although no literature on WW/soil/wheat systems is available, Mordechay and co-workers,[104] among various OMPs, determined EP-CBZ at concentrations up to 1.2 ng/g f.w. in ears of wheat previously irrigated with TWWs of unknown characteristics. Hence, considering a F.I. value of 2.63 g/kg B.W.·d,[105] a D.E. as high as 3.15 ng/kg B.W.·d can be calculated, which is higher than the reference TTC value.

The assessment of the risk of exposure to OMPs described here is undoubtedly limited and, even in the presence of D.E. lower than the corresponding TTC values, does not definitively exclude a hazard for human health deriving from the ingestion of fruit and vegetables produced by irrigation with TWWs. A more accurate assessment of the risk of exposure due to OMPs ingestion would in fact require real toxicity data of the OMPs and their main degradation products, possibly accompanied by equivalent toxicity values with respect to the most toxic compound.

References

1. Nicholson SE, Funk C, Fink AH. Rainfall over the African continent from the 19th through the 21st century. *Global Planet Change* 2018;**165**:114–27.
2. Zittis G. Observed rainfall trends and precipitation uncertainty in the vicinity of the Mediterranean, Middle East and North Africa. *Theor Appl Climatol* 2018;**134**(3–4): 1207–1230.
3. WWAP. *The United Nations world water development report 2017. Wastewater: the untapped resource*. Paris: UNESCO; 2017.
4. Helmecke M, Fries E, Schulte C. Regulating water reuse for agricultural irrigation: risks related to organic micro-contaminants. *Environ Sci Eur* 2020;**32**(1):4.
5. Voulvoulis N. Water reuse from a circular economy perspective and potential risks from an unregulated approach. *Curr Opin Environ Sci Health* 2018;**2**:32–45.
6. Becerra-Castro C, Lopes AR, Vaz-Moreira I, Silva EF, Manaia CM, Nunes OC. Wastewater reuse in irrigation: a microbiological perspective on implications in soil fertility and human and environmental health. *Environ Int* 2015;**75**:117–35.
7. World Health Organization. *WHO guidelines for the safe use of wastewater excreta and greywater*. World Health Organization; 2006.
8. Luo Y, Guo W, Ngo HH, Nghiem LD, Hai FI, Zhang J, et al. A review on the occurrence of micropollutants in the aquatic environment and their fate and removal during wastewater treatment. *Sci Total Environ* 2014;**473**:619–41.
9. Cwiertny DM, Snyder SA, Schlenk D, Kolodziej EP. *Environmental designer drugs: when transformation may not eliminate risk*. ACS Publications; 2014.

10. *Proposal for a regulation of the European parliament and of the council: On minimum requirements for water reuse. COM/2018/337 final—2018/0169.* 2018.
11. Alcalde-Sanz L, Gawlik BM. *Minimum quality requirements for water reuse in agricultural irrigation and aquifer recharge-towards a legal instrument on water reuse at EU level.* Joint Research Centre (European Commission); 2017.
12. *Commission recommendation of 11 September 2014 amending the annex to recommendation 2013/711/EU on the reduction of the presence of dioxins, furans and PCBs in feed and food.* 2014.
13. Wu X, Dodgen LK, Conkle JL, Gan J. Plant uptake of pharmaceutical and personal care products from recycled water and biosolids: a review. *Sci Total Environ* 2015;**536**: 655–666.
14. Verlicchi P, Al Aukidy M, Zambello E. Occurrence of pharmaceutical compounds in urban wastewater: removal, mass load and environmental risk after a secondary treatment—a review. *Sci Total Environ* 2012;**429**:123–55.
15. Xue M, Wu H, Liu S, Huang X, Jin Q, Ren R. Simultaneous determination of 44 pharmaceutically active compounds in water samples using solid-phase extraction coupled with ultra-performance liquid chromatography-tandem mass spectrometry. *Anal Bioanal Chem* 2020;**412**(1):203–22.
16. Archer E, Petrie B, Kasprzyk-Hordern B, Wolfaardt GM. The fate of pharmaceuticals and personal care products (PPCPs), endocrine disrupting contaminants (EDCs), metabolites and illicit drugs in a WWTW and environmental waters. *Chemosphere* 2017; **174**:437–46.
17. Verlicchi P, Al Aukidy M, Jelic A, Petrović M, Barceló D. Comparison of measured and predicted concentrations of selected pharmaceuticals in wastewater and surface water: a case study of a catchment area in the Po Valley (Italy). *Sci Total Environ* 2014;**470**:844–54.
18. Ebele AJ, Abdallah MA-E, Harrad S. Pharmaceuticals and personal care products (PPCPs) in the freshwater aquatic environment. *Emerg Contam* 2017;**3**(1):1–16.
19. Dotan P, Godinger T, Odeh W, Groisman L, Al-Khateeb N, Rabbo AA, et al. Occurrence and fate of endocrine disrupting compounds in wastewater treatment plants in Israel and the Palestinian West Bank. *Chemosphere* 2016;**155**:86–93.
20. Montesdeoca-Esponda S, Checchini L, Del Bubba M, Sosa-Ferrera Z, Santana-Rodriguez JJ. Analytical approaches for the determination of personal care products and evaluation of their occurrence in marine organisms. *Sci Total Environ* 2018;**633**: 405–425.
21. Christou A, Agüera A, Bayona JM, Cytryn E, Fotopoulos V, Lambropoulou D, et al. The potential implications of reclaimed wastewater reuse for irrigation on the agricultural environment: the knowns and unknowns of the fate of antibiotics and antibiotic resistant bacteria and resistance genes—a review. *Water Res* 2017;**123**: 448–467.
22. Liu X, Liang C, Liu X, Zhao F, Han C. Occurrence and human health risk assessment of pharmaceuticals and personal care products in real agricultural systems with long-term reclaimed wastewater irrigation in Beijing, China. *Ecotoxicol Environ Saf* 2020;**190**:110022.
23. Delli Compagni R, Gabrielli M, Polesel F, Turolla A, Trapp S, Vezzaro L, et al. Risk assessment of contaminants of emerging concern in the context of wastewater reuse for irrigation: an integrated modelling approach. *Chemosphere* 2020;**242**:125185.
24. Fatta-Kassinos D, Kalavrouziotis IK, Koukoulakis PH, Vasquez M. The risks associated with wastewater reuse and xenobiotics in the agroecological environment. *Sci Total Environ* 2011;**409**(19):3555–63.
25. Advanced Chemistry Development (ACD/Labs). *Software V11.02 (© 1994–2020 ACD/Labs).* 2020 [Internet]. 2020 [cited 18 June 2020].

26. Sato T, Qadir M, Yamamoto S, Endo T, Zahoor A. Global, regional, and country level need for data on wastewater generation, treatment, and use. *Agric Water Manag* 2013; **130**:1–13.
27. Evgenidou EN, Konstantinou IK, Lambropoulou DA. Occurrence and removal of transformation products of PPCPs and illicit drugs in wastewaters: a review. *Sci Total Environ* 2015;**505**:905–26.
28. Krzeminski P, Tomei MC, Karaolia P, Langenhoff A, Almeida CMR, Felis E, et al. Performance of secondary wastewater treatment methods for the removal of contaminants of emerging concern implicated in crop uptake and antibiotic resistance spread: a review. *Sci Total Environ* 2019;**648**:1052–81.
29. Berardi C, Fibbi D, Coppini E, Renai L, Caprini C, Scordo CVA, et al. Removal efficiency and mass balance of polycyclic aromatic hydrocarbons, phthalates, ethoxylated alkylphenols and alkylphenols in a mixed textile-domestic wastewater treatment plant. *Sci Total Environ* 2019;**674**:36–48.
30. Laht M, Karkman A, Voolaid V, Ritz C, Tenson T, Virta M, et al. Abundances of tetracycline, sulphonamide and beta-lactam antibiotic resistance genes in conventional wastewater treatment plants (WWTPs) with different waste load. *PLoS One* 2014;**9**(8): e103705.
31. Jaramillo MF, Restrepo I. Wastewater reuse in agriculture: a review about its limitations and benefits. *Sustainability* 2017;**9**(10):1734.
32. Richardson SD, Ternes TA. Water analysis: emerging contaminants and current issues. *Anal Chem* 2018;**90**(1):398–428.
33. Wang J, Wang S. Removal of pharmaceuticals and personal care products (PPCPs) from wastewater: a review. *J Environ Manage* 2016;**182**:620–40.
34. Kot-Wasik A, Jakimska A, Śliwka-Kaszyńska M. Occurrence and seasonal variations of 25 pharmaceutical residues in wastewater and drinking water treatment plants. *Environ Monit Assess* 2016;**188**(12):661.
35. Gogoi A, Mazumder P, Tyagi VK, Chaminda GT, An AK, Kumar M. Occurrence and fate of emerging contaminants in water environment: a review. *Groundw Sustain Dev* 2018;**6**:169–80.
36. Illés E, Szabó E, Takács E, Wojnárovits L, Dombi A, Gajda-Schrantz K. Ketoprofen removal by O3 and O3/UV processes: kinetics, transformation products and ecotoxicity. *Sci Total Environ* 2014;**472**:178–84.
37. Christou A, Karaolia P, Hapeshi E, Michael C, Fatta-Kassinos D. Long-term wastewater irrigation of vegetables in real agricultural systems: concentration of pharmaceuticals in soil, uptake and bioaccumulation in tomato fruits and human health risk assessment. *Water Res* 2017;**109**:24–34.
38. Grossberger A, Hadar Y, Borch T, Chefetz B. Biodegradability of pharmaceutical compounds in agricultural soils irrigated with treated wastewater. *Environ Pollut* 2014;**185**:168–77.
39. Malchi T, Maor Y, Tadmor G, Shenker M, Chefetz B. Irrigation of root vegetables with treated wastewater: evaluating uptake of pharmaceuticals and the associated human health risks. *Environ Sci Technol* 2014;**48**(16):9325–33.
40. Calderón-Preciado D, Matamoros V, Savé R, Muñoz P, Biel C, Bayona J. Uptake of microcontaminants by crops irrigated with reclaimed water and groundwater under real field greenhouse conditions. *Environ Sci Pollut Res* 2013;**20**(6):3629–38.
41. Vallecillos L, Borrull F, Pocurull E. An automated headspace solid-phase microextraction followed by gas chromatography–mass spectrometry method to determine macrocyclic musk fragrances in wastewater samples. *Anal Bioanal Chem* 2013;**405**(29):9547–54.
42. Miege C, Choubert J, Ribeiro L, Eusèbe M, Coquery M. Fate of pharmaceuticals and personal care products in wastewater treatment plants—conception of a database and first results. *Environ Pollut* 2009;**157**(5):1721–6.

43. Blaine AC, Rich CD, Sedlacko EM, Hyland KC, Stushnoff C, Dickenson ER, et al. Perfluoroalkyl acid uptake in lettuce (Lactuca sativa) and strawberry (Fragaria ananassa) irrigated with reclaimed water. *Environ Sci Technol* 2014;**48**(24):14361–8.
44. Ciofi L, Renai L, Rossini D, Ancillotti C, Falai A, Fibbi D, et al. Applicability of the direct injection liquid chromatographic tandem mass spectrometric analytical approach to the sub-ng L^{-1} determination of perfluoro-alkyl acids in waste, surface, ground and drinking water samples. *Talanta* 2018;**176**:412–21.
45. Arvaniti OS, Stasinakis AS. Review on the occurrence, fate and removal of perfluorinated compounds during wastewater treatment. *Sci Total Environ* 2015;**524**:81–92.
46. Picó Y, Alvarez-Ruiz R, Alfarhan AH, El-Sheikh MA, Alobaid SM, Barceló D. Uptake and accumulation of emerging contaminants in soil and plant treated with wastewater under real-world environmental conditions in the Al Hayer area (Saudi Arabia). *Sci Total Environ* 2019;**652**:562–72.
47. Paz A, Tadmor G, Malchi T, Blotevogel J, Borch T, Polubesova T, et al. Fate of carbamazepine, its metabolites, and lamotrigine in soils irrigated with reclaimed wastewater: sorption, leaching and plant uptake. *Chemosphere* 2016;**160**:22–9.
48. Tozzi F, Del Bubba M, Petrucci WA, Pecchioli S, Macci C, García FH, et al. Use of a remediated dredged marine sediment as a substrate for food crop cultivation: sediment characterization and assessment of fruit safety and quality using strawberry (Fragaria x ananassa Duch.) as model species of contamination transfer. *Chemosphere* 2020;**238**: 124651.
49. Li H, Qu R, Yan L, Guo W, Ma Y. Field study on the uptake and translocation of PBDEs by wheat (Triticum aestivum L.) in soils amended with sewage sludge. *Chemosphere* 2015;**123**:87–92.
50. Fu Q, Wu X, Ye Q, Ernst F, Gan J. Biosolids inhibit bioavailability and plant uptake of triclosan and triclocarban. *Water Res* 2016;**102**:117–24.
51. Wu C, Spongberg AL, Witter JD, Fang M, Czajkowski KP. Uptake of pharmaceutical and personal care products by soybean plants from soils applied with biosolids and irrigated with contaminated water. *Environ Sci Technol* 2010;**44**(16):6157–61.
52. Shon H, Vigneswaran S, Snyder SA. Effluent organic matter (EfOM) in wastewater: constituents, effects, and treatment. *Crit Rev Environ Sci Technol* 2006;**36**(4):327–74.
53. Graber ER, Gerstl Z. Organic micro-contaminant sorption, transport, accumulation, and root uptake in the soil-plant continuum as a result of irrigation with treated wastewater. *Israel J Plant Sci* 2011;**59**(2–4):105–14.
54. Chefetz B, Mualem T, Ben-Ari J. Sorption and mobility of pharmaceutical compounds in soil irrigated with reclaimed wastewater. *Chemosphere* 2008;**73**(8):1335–43.
55. Navon R, Hernandez-Ruiz S, Chorover J, Chefetz B. Interactions of carbamazepine in soil: effects of dissolved organic matter. *J Environ Qual* 2011;**40**(3):942–8.
56. Thiele-Bruhn S, Seibicke T, Schulten H-R, Leinweber P. Sorption of sulfonamide pharmaceutical antibiotics on whole soils and particle-size fractions. *J Environ Qual* 2004;**33**(4):1331–42.
57. Haham H, Oren A, Chefetz B. Insight into the role of dissolved organic matter in sorption of sulfapyridine by semiarid soils. *Environ Sci Technol* 2012;**46**(21):11870–7.
58. Bernier M-H, Levy GJ, Fine P, Borisover M. Organic matter composition in soils irrigated with treated wastewater: FT-IR spectroscopic analysis of bulk soil samples. *Geoderma* 2013;**209**:233–40.
59. Xu J, Wu L, Chang AC. Degradation and adsorption of selected pharmaceuticals and personal care products (PPCPs) in agricultural soils. *Chemosphere* 2009;**77**(10): 1299–1305.
60. Kodešová R, Grabic R, Kočárek M, Klement A, Golovko O, Fér M, et al. Pharmaceuticals' sorptions relative to properties of thirteen different soils. *Sci Total Environ* 2015;**511**:435–43.

61. Kurwadkar ST, Adams CD, Meyer MT, Kolpin DW. Effects of sorbate speciation on sorption of selected sulfonamides in three loamy soils. *J Agric Food Chem* 2007;**55**(4): 1370–1376.
62. Qiang Z, Adams C. Potentiometric determination of acid dissociation constants (pKa) for human and veterinary antibiotics. *Water Res* 2004;**38**(12):2874–90.
63. Zhang T, Wu B, Sun N, Ye Y, Chen H. Sorption and degradation of wastewater-associated pharmaceuticals and personal care products in agricultural soils and sediment. *Water Sci Technol* 2013;**68**(5):991–8.
64. Fischer T. Humic supramolecular structures have polar surfaces and unpolar cores in native soil. *Chemosphere* 2017;**183**:437–43.
65. Durán-Álvarez JC, Prado-Pano B, Jiménez-Cisneros B. Sorption and desorption of carbamazepine, naproxen and triclosan in a soil irrigated with raw wastewater: estimation of the sorption parameters by considering the initial mass of the compounds in the soil. *Chemosphere* 2012;**88**(1):84–90.
66. Gibson R, Durán-Álvarez JC, Estrada KL, Chávez A, Cisneros BJ. Accumulation and leaching potential of some pharmaceuticals and potential endocrine disruptors in soils irrigated with wastewater in the Tula Valley, Mexico. *Chemosphere* 2010;**81**(11):1437–45.
67. Qin Q, Chen X, Zhuang J. The fate and impact of pharmaceuticals and personal care products in agricultural soils irrigated with reclaimed water. *Crit Rev Environ Sci Technol* 2015;**45**(13):1379–408.
68. Verlicchi P, Zambello E. Pharmaceuticals and personal care products in untreated and treated sewage sludge: occurrence and environmental risk in the case of application on soil—a critical review. *Sci Total Environ* 2015;**538**:750–67.
69. Styszko K. Sorption of emerging organic micropollutants onto fine sediments in a water supply dam reservoir, Poland. *J Soil Sediment* 2016;**16**(2):677–86.
70. Amini S, Ghadiri H, Chen C, Marschner P. Salt-affected soils, reclamation, carbon dynamics, and biochar: a review. *J Soil Sediment* 2016;**16**(3):939–53.
71. Chaganti VN, Ganjegunte G, Niu G, Ulery A, Flynn R, Enciso JM, et al. Effects of treated urban wastewater irrigation on bioenergy sorghum and soil quality. *Agric Water Manag* 2020;**228**:105894.
72. Salgado-Méndez S, Gilabert-Alarcón C, Daesslé L, Mendoza-Espinosa L, Avilés-Marín S, Stumpp C. Short-term effects on agricultural soils irrigated with reclaimed water in Baja California, México. *Bull Environ Contam Toxicol* 2019;**102**(6):829–35.
73. Yu Y, Liu Y, Wu L. Sorption and degradation of pharmaceuticals and personal care products (PPCPs) in soils. *Environ Sci Pollut Res* 2013;**20**(6):4261–7.
74. Al-Farsi RS, Ahmed M, Al-Busaidi A, Choudri B. Translocation of pharmaceuticals and personal care products (PPCPs) into plant tissues: a review. *Emerg Contam* 2017;**3**(4): 132–137.
75. Bartrons M, Peñuelas J. Pharmaceuticals and personal-care products in plants. *Trends Plant Sci* 2017;**22**(3):194–203.
76. Pullagurala VLR, Rawat S, Adisa IO, Hernandez-Viezcas JA, Peralta-Videa JR, Gardea-Torresdey JL. Plant uptake and translocation of contaminants of emerging concern in soil. *Sci Total Environ* 2018;**636**:1585–96.
77. Olowoyo J, Mugivhisa L. Evidence of uptake of different pollutants in plants harvested from soil treated and fertilized with organic materials as source of soil nutrients from developing countries. *Chem Biol Technol Agric* 2019;**6**(1):1–11.
78. Al Nasir F, Batarseh MI. Agricultural reuse of reclaimed water and uptake of organic compounds: pilot study at Mutah University wastewater treatment plant, Jordan. *Chemosphere* 2008;**72**(8):1203–14.
79. Beltrán EM, Pablos MV, Torija CF, Porcel MÁ, González-Doncel M. Uptake of atenolol, carbamazepine and triclosan by crops irrigated with reclaimed water in a Mediterranean scenario. *Ecotoxicol Environ Saf* 2020,**191**.110171.

80. Goldstein M, Shenker M, Chefetz B. Insights into the uptake processes of wastewater-borne pharmaceuticals by vegetables. *Environ Sci Technol* 2014;**48**(10):5593–600.
81. Li Y, Huang G, Gu H, Huang Q, Lou C, Zhang L, et al. Assessing the risk of phthalate ester (PAE) contamination in soils and crops irrigated with treated sewage effluent. *Water* 2018;**10**(8):999.
82. Martínez-Piernas AB, Plaza-Bolaños P, Fernández-Ibáñez P, Agüera A. Organic microcontaminants in tomato crops irrigated with reclaimed water grown under field conditions: occurrence, uptake, and health risk assessment. *J Agric Food Chem* 2019;**67**(25): 6930–6939.
83. Pan M, Wong CK, Chu L. Distribution of antibiotics in wastewater-irrigated soils and their accumulation in vegetable crops in the Pearl River Delta, Southern China. *J Agric Food Chem* 2014;**62**(46):11062–9.
84. Li Y, Sallach JB, Zhang W, Boyd SA, Li H. Insight into the distribution of pharmaceuticals in soil-water-plant systems. *Water Res* 2019;**152**:38–46.
85. Feng NX, Yu J, Zhao HM, Cheng YT, Mo CH, Cai QY, et al. Efficient phytoremediation of organic contaminants in soils using plant-endophyte partnerships. *Sci Total Environ* 2017;**583**:352–68.
86. Riemenschneider C, Al-Raggad M, Moeder M, Seiwert B, Salameh E, Reemtsma T. Pharmaceuticals, their metabolites, and other polar pollutants in field-grown vegetables irrigated with treated municipal wastewater. *J Agric Food Chem* 2016;**64**(29):5784–92.
87. Pan M, Chu L. Transfer of antibiotics from wastewater or animal manure to soil and edible crops. *Environ Pollut* 2017;**231**:829–36.
88. Hyland KC, Blaine AC, Higgins CP. Accumulation of contaminants of emerging concern in food crops—part 2: plant distribution. *Environ Toxicol Chem* 2015;**34**(10): 2222–2230.
89. Scordo CVA, Checchini L, Renai L, Orlandini S, Bruzzoniti MC, Fibbi D, et al. Optimization and validation of a method based on QuEChERS extraction and liquid chromatographic-tandem mass spectrometric analysis for the determination of perfluoroalkyl acids in strawberry and olive fruits, as model crops with different matrix characteristics. *J Chromatogr A* 2020;**1621**:461038.
90. Covello VT, Merkhoher MW. *Risk assessment methods: approaches for assessing health and environmental risks*. Springer Science & Business Media; 2013.
91. Prosser R, Sibley P. Human health risk assessment of pharmaceuticals and personal care products in plant tissue due to biosolids and manure amendments, and wastewater irrigation. *Environ Int* 2015;**75**:223–33.
92. World Health Organization. *Principles and methods for the risk assessment of chemicals in food*. World Health Organization; 2009.
93. USEPA. Chapter 9: intake of fruits and vegetables (updated in 2018). In: *Exposure factors handbook: 2011 edition. EPA/600/R-09/052F*. Washington, DC: USEPA, Office of Research and Development; 2011.
94. Aryal N, Reinhold DM. Phytoaccumulation of antimicrobials from biosolids: impacts on environmental fate and relevance to human exposure. *Water Res* 2011;**45**(17): 5545–5552.
95. Duri LG, Fiorentino N, Cozzolino E, Ottaiano L, Agrelli D, Fagnano M. Bioassays for evaluation of sanitary risks from food crops cultivated in potentially contaminated sites. *Ital J Agron* 2018;**13**(s1):45–52.
96. *Foodstuff composition tables*. Accessed on 15 May 2020 [Internet]. Last update: December 2019. Available from https://www.alimentinutrizione.it/tabelle-nutrizionali/005730.
97. Dorne JL, Richardson J, Kass G, Georgiadis N, Monguidi M, Pasinato L, et al. OpenFoodTox: EFSA's open source toxicological database on chemical hazards in food and feed. *EFSA J* 2017;**15**(1):e15011.

98. Blackburn K, Stickney JA, Carlson-Lynch HL, McGinnis PM, Chappell L, Felter SP. Application of the threshold of toxicological concern approach to ingredients in personal and household care products. *Regul Toxicol Pharmacol* 2005;**43**(3):249–59.
99. EFSA. Scientific opinion on exploring options for providing advice about possible human health risks based on the concept of threshold of toxicological concern (TTC). *EFSA J* 2012;**10**(7):2750.
100. Kroes R, Renwick A, Cheeseman M, Kleiner J, Mangelsdorf I, Piersma A, et al. Structure-based thresholds of toxicological concern (TTC): guidance for application to substances present at low levels in the diet. *Food Chem Toxicol* 2004;**42**(1):65–83.
101. EFSA. Guidance on the use of the threshold of toxicological concern approach in food safety assessment. *EFSA J* 2019;**17**(6):e05708.
102. Reilly L, Serafimova R, Partosch F, Gundert-Remy U, Abrahantes JC, Dorne J-LM, et al. Testing the thresholds of toxicological concern values using a new database for food-related substances. *Toxicol Lett* 2019;**314**:117–23.
103. Cramer G, Ford R, Hall R. Estimation of toxic hazard—a decision tree approach. *Food Cosmet Toxicol* 1976;**16**(3):255–76.
104. Mordechay EB, Tarchitzky J, Chen Y, Shenker M, Chefetz B. Composted biosolids and treated wastewater as sources of pharmaceuticals and personal care products for plant uptake: a case study with carbamazepine. *Environ Pollut* 2018;**232**:164–72.
105. *Global wheat per capita consumption.* Accessed on 15 May 2020 [Internet]. Last update: 2019. Available from http://statistics.amis-outlook.org/data/#COMPARE.

CHAPTER FIVE

Environmental risks of sewage sludge reuse in agriculture

Elisabet Marti[a], Victoria Osorio[b], Marta Llorca[c], Lidia Paredes[b], and Meritxell Gros[d,e,*]

[a]Food Microbial Systems, Agroscope, Bern, Switzerland
[b]Beta Technological Centre (TECNIO Network), University of Vic-Central University of Catalonia, Barcelona, Spain
[c]Institute of Environmental Assessment and Water Research (IDAEA-CSIC), Barcelona, Spain
[d]Catalan Institute for Water Research (ICRA), Girona, Spain
[e]University of Girona (UdG), Girona, Spain
*Corresponding author: e-mail address: mgros@icra.cat

Contents

1. Introduction 138
2. Overview of CECs in sewage sludge 139
3. Environmental risks associated with sewage sludge reuse 143
   3.1 Occurrence of CECs in soils 144
   3.2 Ecotoxicological effects and risks in soils 147
   3.3 Transport in soil and contamination of natural water resources 151
   3.4 Plant/crop uptake 153
   3.5 Phytotoxic effects 158
   3.6 Public health implications 164
   3.7 Dissemination of antibiotic resistance 165
   3.8 Strategies to reduce environmental risks 167
4. Conclusions 168
Acknowledgments 169
References 169

Abstract

Sewage sludge reuse will become crucial for the sustainability of agriculture, given the increasing population and the lack of essential nutrients for crop growth. Nevertheless, this practice raises some concerns, due to the large amounts of contaminants of emerging concern (CECs) present in sewage sludge, including pharmaceuticals and personal care products (PPCPs), microplastics (MPs), engineered nanomaterials (ENMs), *per*- and polyfluoroalkyl substances (PFASs), flame retardants, plasticizers, metals, antibiotic resistance bacteria (ARB) and resistance genes (ARGs). These CECs may adversely affect the environment and, ultimately, human health through the consumption of crops grown under sludge-amended soils. This chapter reviews the potential environmental and human health risks associated with sewage sludge reuse in agriculture, by focusing

Advances in Chemical Pollution, Environmental Management and Protection, Volume 6
ISSN 2468-9289
https://doi.org/10.1016/bs.apmp.2020.07.003

on relevant classes of CECs, such as PPCPs, MPs, carbon-based ENMs and considering ARBs and ARGs, a topic so far poorly addressed. The risks discussed focus on the contamination of aquatic and terrestrial ecosystems and on CECs uptake by crops intended for human consumption. Finally, technological strategies to improve sludge treatment, and thus minimize such risks, are briefly discussed.

Keywords: Sewage sludge reuse, Contaminants of emerging concern, Antibiotic resistance, Soil and water pollution, Plant uptake

1. Introduction

The application of sewage sludge as fertilizer in crops is already a common way for disposal in the USA, the UK, Germany and southern Europe.[1] Indeed, in the European Union (EU), approximately 50% of the total sludge produced is spread on agricultural soils directly or after treatment.[2] The agricultural European Innovation Partnership (EIP-AGRI) has already warned that European agriculture needs to progress toward more closed loops concerning the provision of nutrients.[3] Therefore, the use of sewage sludge as a bio-fertilizer represents a worthy practice for recirculating valuable nutrients, such as phosphorus and nitrogen, to food chains via crop plants.[4] Nevertheless, the safety of this practice is constantly under discussion, given the large amounts of contaminants of emerging concern (CECs) present in sewage sludge, such as pharmaceuticals and personal care products (PPCPs), microplastics (MPs), engineered nanomaterials (ENMs), *per-* and polyfluoroalkyl substances (PFASs), flame retardants, plasticizers, metals, pathogens, antibiotic resistance bacteria (ARB) and resistance genes (ARGs), that may adversely affect the environment and, ultimately, human health.

In the EU, the land application of sewage sludge started being regulated in 1986 by Council Directive 86/278/EEC, although every single country has its own sludge management system governed by internal laws. Current legislations that control the spread of sewage sludge onto arable land, set-up limits for heavy metals and pathogens,[5] while no guideline values are still available for CECs. In order to decrease the potential environmental hazards associated with sewage sludge reuse, several treatments are applied to convert it into safer biosolids.[6] The most common treatments are aerobic and anaerobic digestion (AD), dewatering, heating, composting, gravity thickening or lime-stabilization.[7] AD is the most used method to obtain biosolids but there is controversial data regarding the elimination of CECs, human pathogens, ARB and ARGs during these processes.[8]

In this context, this chapter reviews the potential environmental risks associated with sewage sludge reuse in agriculture and discusses the most novel technological strategies to enhance CECs removal and thus, minimize risks. Among the large variety of CECs present in sewage sludge, this chapter will focus on selected groups of contaminants such as: (i) PPCPs, based on their potential to exert ecotoxicological and human health risks, (ii) MPs and (iii) carbon-based ENMs, because they are considered as one of the most novel classes of CECs and the body of literature discussing their risks associated with sewage sludge reuse is very limited. The potential spread of antimicrobial resistance (AMR) through sewage sludge reuse will be also discussed.

2. Overview of CECs in sewage sludge

Several studies reported on the occurrence of PPCPs in sewage sludge.[9,10] Fig. 1 summarizes the range of minimum and maximum concentrations detected in untreated sewage sludge for some of the most representative compounds, including antibiotics (sulfamethazine, sulfamethoxazole, ciprofloxacin, ofloxacin, oxytetracycline, tetracycline, azithromycin and erythromycin), analgesics and anti-inflammatories (ibuprofen and naproxen),

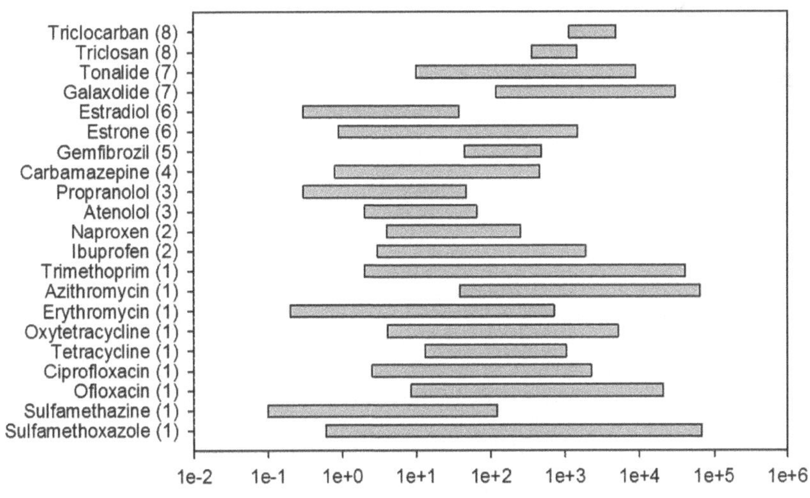

Fig. 1 Range of concentrations (ng/g DM) of representative PPCPs in sewage sludge. (1) Antibiotics; (2) analgesics and anti-inflammatories; (3) β-blockers; (4) psychiatric drugs; (5) lipid regulators, (6) hormones; (7) musk fragrances and (8) antiseptics.

β-blockers (atenolol and propranolol), psychiatric drugs (carbamazepine), lipid regulators (gemfibrozil), hormones (estradiol and estrone), musk fragrances (tonalide and galaxolide) and antiseptics (triclocarban and triclosan). Concentration ranges detected cover several orders of magnitude, from below 1 ng/g DM (dry matter) up to almost 10^5 ng/g DM for some compounds. Antibiotics (sulfamethoxazole, azithromycin, trimethoprim, ofloxacin) and musk fragrances (galaxolide) are the substances showing the highest maximum concentrations (10^4–10^5 ng/g DM). Data in the scientific literature shows that the levels of PPCPs found in digested and composted sludge are somewhat lower (with maximum concentrations between the 10^2 and 10^3 ng/g DM range). Nevertheless, some of the selected PPCPs are still present at the 10^4 ng/g DM concentration levels, such as musk fragrances (galaxolide) and the antiseptic triclosan.[10]

MPs, and nanoplastics (NPs), detected in wastewater treatment plants (WWTPs), and thus in sludge, are smaller in size and have different appearance and profiles than the plastic commonly observed in the environment.[11] MPs reach WWTP mostly from clothing and microplastic beads from personal care products and cosmetics.[11–13] Some studies reported that between 80% and 90% of MPs are trapped in sewage sludge during wastewater treatment[14–18] and that most MPs observed in sludge are fibers.[16,19] The other part of MPs include microparticles (with a ratio fiber/microparticle higher than 0.9 in some cases[14]). From those microparticles, the majority are composed by films (58% of the non-fiber particles with median surface area of 0.5 mm[2,14]), beads, fragments and glitter plastics.[20] Regarding polymer type, Microscope Fourier Transform Infrared-spectroscopy revealed that most fibers in sludge belong to polyolefin, acrylic fibers, polyethylene and polyamide[16] while microparticles are made of polyethylene, polyethylene terephthalate, polyacrylonitrile, polypropylene, polystyrene and polyamide.[21,22] Nonetheless, some authors observed different sludge-based MPs contents depending on the sampling months within the same WWTP. For example, Li et al.[16] observed variations in China, with western facilities having less MPs contents than eastern facilities. In terms of the amounts of MPs detected, they generally range from 1 to 100×10^3 particles/kg of dry sludge and the particles are from 0.7 μm to 5 mm in most of the cases (Table 1).

ENMs are manmade carbon-based or inorganic materials having particle sizes <100 nm,[27] that are used in many applications, such as in medicine, agriculture and in personal care products.[28] Within ENMs, carbon-based engineered nanomaterials (CBENMs) deserve special attention due to their potential toxicity to humans and the environment.[29] CBENMs include

Table 1 Number of microplastic in sewage sludge and soil.

	Site	MP size range	MP amounts	References
Biosolid from secondary WWTP	Southern California (USA)	<5 mm	5 particles/5 g	11
Sludge	Chile	>1 mm	18–41 particles/g	14
Sludge WWTP	China	37 μm–5 mm	$1.60–56.4 \times 10^3$ particles/kg dw	16
WWTPs	Germany	10 μm–5 mm	1000–24,000 particles/kg dw	22
Sludge WWTPs	Switzerland	300 μm–5 mm	$16.7 \pm 1.96 \times 10^3$ particles/kg dw	23
WWTPs	Norway	54 μm–4.99 mm	1701–19,837 particles/kg dw	20
Sludge WWTP	USA	No data	1–4 particles/g dw	19
Sludge WWTP	Netherlands	0.7 μm–5 mm	370–950 particles/kg ww	24
Sludge WWTPs	Ireland	250 μm–4 mm	4196–15,385 particles/kg dw	25
Sludge WWTPs	China	20 μm–5 mm	5553–13,460 particles/kg	21
Agricultural surface soil	China	20 μm–5 mm	420–1290 particles/kg	21
Agricultural soil from cropped areas	China	50 μm–1 mm	1080–22,320 particles/kg	26

fullerenes (e.g., C_{60}), carbon nanodots (C-dots), quantum dots (QDs), single-walled carbon nanotubes (SWCNT) and multiwalled carbon nanotubes (MWCNT). Carbon nanotubes (CNTs) are worldwide produced with up to 100–1000 t/year, while the amount of QDs and fullerenes is less than 10 t/year.[30] CBENMs released from commercial products end up in urban wastewater and a remarkable percentage sorb to sewage sludge in WWTPs. For instance, batch experiments demonstrated that up to 94% of aqueous fullerenes are removed from wastewater by adsorption to the biomass in the WWTP.[31] Among these, 83–96% of nC_{60} is sorbed onto sludge.[32,33] Furthermore, according to studies based on predicted values,

Table 2 Concentrations of CBENPs in sewage sludge and soil.

Compounds	Sludge (ng/kg)	Soil (Δng/kg y)	Sludge-treated soil (Δng/kg y)	References
Fullerenes	EU: 12000	EU: 0,058	EU: 2,2	35
	US: 10000	US: 0,024	US: 1,01	
	EU: 90000	EU: 0,10	EU: 620	36
CNTs	EU: 62000	EU: 1,51	EU: 73,6	35
	US: 68000	US: 0,56	US: 31,4	
	EU: 15000	EU: 5,1	EU: 990	36
	EU: 273000	EU: 33,8	EU: 10200	37

concentrations in sewage sludge are estimated to be between 0.0011 and 10 ppm.[34] Estimated concentrations for CBENMs are compiled in Table 2.

Raw sewage sludge may also contain a wide variety of pathogenic microorganisms, including bacteria, viruses, protozoa, helminths and fungi and their occurrences have been widely reviewed.[38,39] Several reports have also proved the presence of ARBs and ARGs in sewage sludge. Indeed, a mass balance analysis revealed that around 90% of the ARGs in WWTPs accumulate into the sludge, while only 10% goes to the treated water effluent.[40] In fact, one study applying quantitative metagenomic and metatranscriptomic approaches concluded that the nitrifying and denitrifying bioreactor sludge had significantly higher transcription values of ARGs (around 10^{11} transcripts/L) in comparison to the other compartments and there was a strong correlation of ARGs with biomass.[41] A research performed at the largest WWTP in Croatia concluded that human-associated carbapenem resistant bacteria reached values of 2.5 log colony forming units/g in the activated sludge, and these amounts remained within the same order in mesophilic-digested sludge. However, when this sludge was stabilized by dewatering and lime treatment these resistant bacteria were eliminated.[42] Regarding ARGs, a study performed in three WWTPs observed that several ARGs conferring resistance to clinically important antibiotics, such as beta-lactams, tetracyclines and aminoglycosides, increased 50-fold their relative abundance in surplus sludge compared to primary sludge.[43] Several studies quantifying ARG in raw sludge by quantitative PCR (qPCR) reported values of 10^{10}–10^{11} copies/g DM in northern[44] and eastern China,[45] mainly for genes conferring resistance to tetracyclines and sulfonamides. Another study from southern China, also demonstrated that tetracycline and sulfonamide resistance genes presented the highest relative abundance (normalized by 16S rRNA concentration) in the raw sewage sludge.[46] Moreover, He and co-authors[45] were able to discriminate free ARGs (eARGs) from intracellular genes and reported concentrations up to 10^{10} copies/g DM of eARGs, demonstrating the persistence of extracellular ARGs and mobile genetic elements (MGE) which can be also mobilized to the competent non-resistant bacteria. Following the same tendency, a study performed in Spain showed that sulfonamide resistance genes were the most prevalent in sewage sludge, reaching 10^8 copies/g, not directly comparable with the studies expressing the concentrations in dry weight.[47] With respect to ARGs in treated sludge, *tet*O and *tet*W (conferring resistance to tetracyclines) and *sul*I genes (conferring resistance to sulfonamides) presented concentrations between 10^7 and 10^{10} copies/g in sludge treated by dewatering, gravity thickening, anaerobic digestion and lime stabilization.[48] Tian and colleagues[49] applied metagenomics to elucidate the abundance of ARGs and showed that thermophilic sludge held a lower total ARG abundance (50.65 ppm) than mesophilic sludge (125.97 ppm), although ARG subtypes exhibited varied responses to mesophilic and thermophilic treatment. Regarding composted sludge, one study reported concentrations of 10^{11} ARGs copies/g dry weight in the initial sludge, being the aminoglycosides the most prevalent ARGs, which were maintained and enriched during composting.[50] Another similar research also reported initial concentrations of ARGs and MGEs of 10^{10}–10^{11} gene copies/g DM, with tetracycline and sulfonamide resistance genes being the dominant genes. In this case, the relative abundances of ARGs decreased significantly in hyperthermophilic composted sludge (0.05 copies/16S rRNA gene) compared to conventional compost (0.14 copies/16S rRNA gene).[51]

3. Environmental risks associated with sewage sludge reuse

The fertilization of agricultural soils with sewage sludge may result in the continuous exposure of the agricultural environment to CECs and to the dissemination of pathogens and antimicrobial resistance[52] (Fig. 2). Once sludge is spread onto arable land, these contaminants may accumulate and persist in soils, leach to surface and groundwater or be taken up by crops, posing potential environmental and human health risks, as discussed in detail in Sections 3.1–3.8.

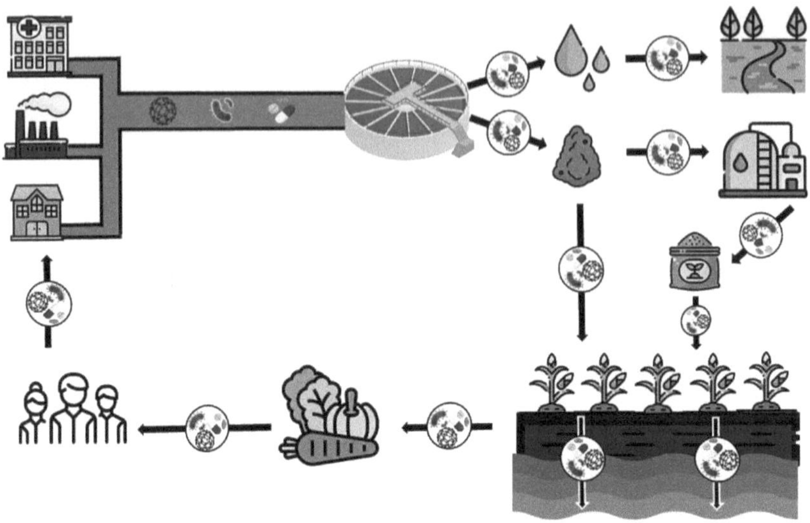

Fig. 2 Environmental risks associated with sewage sludge reuse: the dissemination of CECs and AMR to the environment. *Image created in www.flaticon.com.*

3.1 Occurrence of CECs in soils

Data about the occurrence of PPCPs in sludge-amended soils is so far scarce. A recent study estimated an annual load for a wide range of pharmaceuticals, illicit drugs and their metabolites in agricultural soils following sewage sludge application in the Slovak Republic, assuming no degradation of these compounds.[5] Substances showing the highest maximum input loads into the soils were the anti-histaminic fexofenadine (120 kg/year), the anti-hypertensives verapamil (28 kg/year) and telmisartan (16 kg/year), the psychiatric drug citalopram (21 kg/year) and the anti-inflammatory diclofenac (12 kg/year). Regarding the concentrations of PPCPs detected in sludge-amended soils, Table 3 summarizes the levels found for some representative PPCPs. Fluoroquinolone antibiotics were one of the therapeutic groups detected at the highest concentrations in soil (Table 3). The study performed by Golet et al.[56] found norfloxacin and ciprofloxacin at concentrations ranging from 290 to 400 ng/g DM and from 270 to 300 ng/g DM in topsoil 8 and 21 months after sludge application, respectively. Similar concentrations were found in the study by Walters et al.,[53] who explored the fate of 72 PPCPs in an outdoor mesocosms experiment using sludge-amended soil over the course of 3 years. In this study, fluoroquinolone antibiotics were also present at remarkable concentrations (~50–550 ng/g DM). Nevertheless, the antibacterial agents triclosan and triclocarban were the compounds found at the highest

Table 3 Concentrations of PPCPs in sewage sludge-amended soils.

Compounds	Measured concentrations (MEC) (ng/g DM)	References
Carbamazepine	6 (day 0 in a mesocosms experiment)	53
	<LOQ-0.25 (soil amended with composted sludge)	54
Gemfibrozil	30 (day 0 in a mesocosms experiment)	53
Miconazole	60 (day 0 in a mesocosms experiment)	53
Diphenhydramine	40 (day 0 in a mesocosms experiment)	53
	<LOD-40 (sludge amended field)	55
Fluoxetine	10 (day 0 in a mesocosms experiment)	53
Azithromycin	30 (day 0 in a mesocosms experiment)	53
Ofloxacin	450 (day 0 in a mesocosms experiment)	53
	0.3–8.6 (soil amended with composted sludge)	54
Ciprofloxacin	350–400 after 8 months sludge application	56
	270–280 after 21 months sludge application	56
	450 (2.5 cm depth)	57
	550 (day 0 in a mesocosms experiment)	53
	<LOQ-8.7 (soil amended with composted sludge)	54
Norfloxacin	320–290 after 8 months sludge application	56
	270–300 after 21 months application	56
	350 (2.5 cm depth)	57
	50 (day 0 in a mesocosms experiment)	53
	<LOQ-9.4 (soil amended with composted sludge)	54
Tetracycline	20 (day 0 in a mesocosms experiment)	53
Doxycycline	15 (day 0 in a mesocosms experiment)	53
Triclosan	1200 (day 0 in a mesocosms experiment)	53
Triclocarban	3000 (day 0 in a mesocosms experiment)	53
Galaxolide (HHCB)	1050–2770 (sludge amended field)	55
	0.37–49.9 (sludge amended soil)	58

Continued

Table 3 Concentrations of PPCPs in sewage sludge-amended soils.—cont'd

Compounds	Measured concentrations (MEC) (ng/g DM)	References
Tonalide (AHTN)	287–773 (sludge amended field)	55
	0.37–23.7 (sludge amended soil)	58
Acetophenone	285–540 (sludge amended field)	55
Dienestrol (DNS)	0.98–3.3 (sludge amended soils)	59
Diethylstilbestrol (DES)	<LOD-0.4 (sludge amended soils)	59
Estrone (E1)	0.25–0.41 (sludge amended soils)	59
Mestranol (MES)	0.10–0.21 (sludge amended soils)	59

levels, ranging from approximately 1200 to 3000 ng/g DM (Table 3), while other pharmaceuticals, such as tetracycline, doxycycline, carbamazepine, gemfibrozil, miconazole, diphenhydramine, fluoxetine and azithromycin, were detected at lower levels (from 6 to 60 ng/g DM). The dissipation of PPCPs was also investigated, and few substances, including diphenhydramine, fluoxetine and triclocarban showed no discernible loss over the 3 years monitoring period, while azithromycin, ciprofloxacin and norfloxacin had moderately high half-lives (from 990 to 3466 days), highlighting the persistence of these compounds in sludge-amended soils. Triclosan was less persistent, with lower half-lives from 182 to 193 days, and this coincides with other studies where triclosan was reduced up to 80% after one-year sludge application to soil.[60] High concentrations for musk fragrances were also observed in soils (Table 3). In the study performed by Kinney et al.[55] the occurrence of 77 wastewater derived contaminants, including PPCPs, in sludge and manure amended soils, and their bioaccumulation in earthworms, was assessed, and 25 out of the 77 PPCPs were detected in earthworms in the biosolids-amended site (seven of which exceeded the concentrations of 1000 ng/g DM). Lower PPCPs concentrations were observed in the long-term field experiment conducted by Bourdat-Deschamps et al.[54] In this study, experimental plots were fertilized with different organic waste products, including sewage sludge. The most persistent compounds were fluoroquinolone antibiotics and carbamazepine, which were detected at concentrations in the low ng/g DM range.

For MPs, it is estimated that among the total loads arriving to WWTPs, between 125 and 850 tons of MP/million inhabitants are added annually to European agricultural soils either through direct application of sewage sludge or as processed biosolids,[61] and these amounts are greater than

MPs entering the world's ocean.[62] Some studies pointed out an increase of MPs counts in soils where sewage sludge has been applied as fertilizer. For instance, the presence of fiber MPs has been reported in agricultural fields up to 15 years after sewage sludge amendment and they still maintained their original properties.[19] In a more recent study, Corradini et al.[14] evidenced the accumulation of MPs over time in agricultural fields with different sludge application records. In general, the particle types detected are in accordance with those identified in sewage sludge and fibers are the most abundant MPs types in soils[14,19,26,61] followed by films and fragments (less than 10% in samples).[26] In terms of amounts detected, the most usual values are between 400 and 10^4 particles/kg dry soil (Table 1), being fibers the most dominating morphology with sizes between 0.02 and 0.25 mm.[21,26] Finally, the composition of those MPs is dominated by the presence of polyethylene and polypropylene,[21] two of the most used plastics.

For CBENMs, data available on measured concentrations in soils is scarce due to the lack of reliable analytical methodologies able to detect and quantify them.[63,64] Since CBENMs occur as colloidal systems, the conditions applied during the analytical protocols may change their physicochemical properties, potentially triggering artifacts during analysis.[65] Moreover, CBENMs are present in such low concentrations that cannot be differentiated from the background signal produced by particulate materials, including natural nanoparticles, which are always concomitant in an environmental matrix.[66–68] For this reason, most studies dealing with the occurrence and fate of CBENMs in soils are based on predicted environmental concentrations (PECs) by using different modeling approaches. Table 2 summarizes some PECs estimated for fullerenes and carbon nanotubes (CNTs). For instance, Gottschall and co-authors[69] estimated that the application of sewage sludge in Europe containing 12,000 ng/kg of fullerenes would result in an annual increase in concentration from 0,058 ng/kg present in soils to 2.2 ng/kg in sludge-treated soils (Table 2). The increase in concentrations in sludge-amended soils is even greater for CNTs, where concentrations of 10,200 ng/kg were predicted to be detected in sludge-amended soils in Europe (Table 2). Other studies estimated that PECs for CNTs would be up to 10 and 0.5 mg/kg in sediments and soils.[34,70,71]

3.2 Ecotoxicological effects and risks in soils

The occurrence of CECs in soils may exert undesired ecotoxicological effects. PPCPs, and especially antibiotics, are likely to disturb the complex regulatory networks in the soil microbiome and soil fauna, which are closely related with soil quality and ecological function.[72] Indeed, the study performed by Pino-Otín et al.[73] demonstrated that the occurrence of PPCPs in soils has an impact on microbial communities by changing the ability of microbes to metabolize different carbon sources, thus affecting metabolic diversity of the soil community. Among the 18 compounds evaluated, non-steroidal anti-inflammatory (NSAIDs) drugs and antibiotics were the least and most toxic, respectively, while lipid regulators and β-blockers presented intermediate toxicity.

Nevertheless, the most common approach to assess the potential ecotoxicological risks posed by the occurrence of PPCPs in the terrestrial environment are based on the estimation of risk or hazard quotients.[74] Risk quotients (RQs) are calculated as the ration between measured (MEC) or predicted environmental concentrations (PEC) in soils with predicted no-effect concentrations ($PNEC_{soil}$). PNECs are usually calculated by using the lowest acute toxicity data (EC_{50} or LC_{50} values) and dividing them with an assessment factor of 1000 to consider the worst-case scenarios. Due to the lack of data on acute toxicity for terrestrial organisms, such as earthworms, plants and soil microorganisms, few studies used these endpoints to estimate PNEC values in soils.[54,75] Instead, most studies relied on aquatic toxicity data to calculate PNECs using three model organisms (fish, *Daphnia magna* and algae), which are afterwards corrected with the compounds solid/water partition coefficient (K_d), the soil bulk density and the field water content.[10,75–77] With the RQ approach, the criteria used to establish whether potential environmental risks may occur are: (a) low risk when RQ <0.1; (b) medium risk with $0.1 < RQ < 1$, and high risk for $RQ \geq 1$.[10,54,76] In the study carried out by Verlicchi et al.[10] the RQ approach was used to estimate the potential risks posed by the PPCPs found in digested sewage sludge once they are spread on agricultural soils. The range of RQs using the minimum and maximum concentrations found in digested sludge are summarized in Fig. 3. Compounds such as antibiotics (sulfamethoxazole, erythromycin, roxithromycin, azithromycin and ofloxacin), hormones (estrone (E1), 17β-estradiol (E2) and ethinylestradiol (EE2)), analgesics and anti-inflammatories (acetaminophen, ibuprofen, naproxen and salicylic acid) and the β-blocker propranolol were those highlighted to pose the highest environmental risks. Most of these compounds were also identified as the most relevant in other studies. For instance, Martín et al.[76,77] assessed the risks posed by the PPCPs found in digested and composted sewage sludge and sludge-amended soils and the hormones 17β-estradiol and 17α-ethinylestradiol, the anti-inflammatory ibuprofen, the lipid regulator gemfibrozil and the

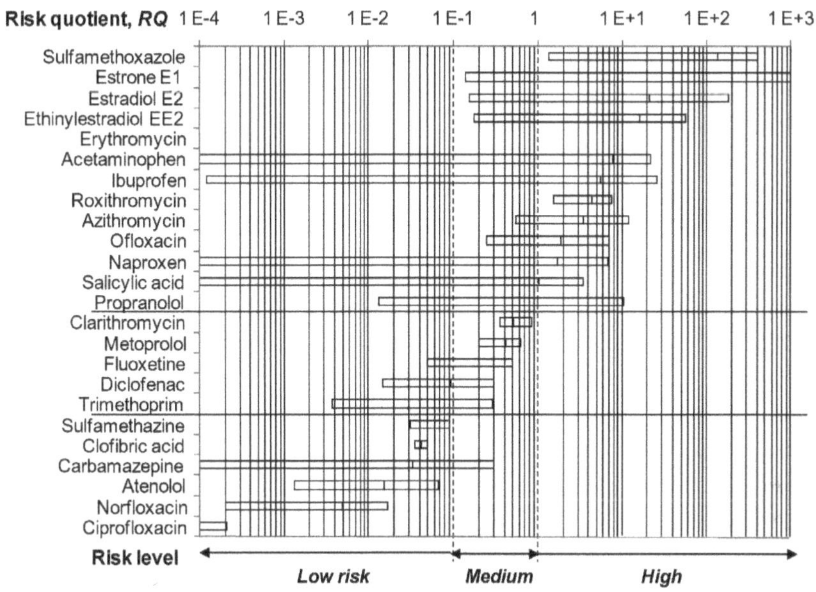

Fig. 3 Risk quotients of representative PPCPs in digested sewage sludge. *Taken from Verlicchi P, Zambello E. Pharmaceuticals and personal care products in untreated and treated sewage sludge: occurrence and environmental risk in the case of application on soil—a critical review.* Sci Total Environ *2015;**538**:750–67.*

antibiotic sulfamethoxazole were pointed out as the most relevant compounds as well. As expected, a drastic decrease in RQ was observed after sludge application in soils, being 17β-estradiol the only substance showing some potential toxic effects. In the study conducted by Thomaidi et al.[75] the antiseptic triclosan was also highlighted as a relevant chemical. In this study, risks due to the occurrence of compound mixtures, and not only those attributed to single compounds were evaluated. Results showed that the estimated risk quotient largely exceeded the threshold of one, indicating that the co-existence of a wide variety of CECs in sewage sludge will result to a series of ecological hazards for terrestrial organisms in sludge-amended soils.

The occurrence of MPs in sludge-amended soils has still unknown impacts for soil quality loss, soil organisms, livestock and human health.[78] Nonetheless, since MPs cannot be (bio)degraded, the main environmental problem is their persistence in the soil environment.[13] The risks of MPs associated with sludge or biosolids reuse as fertilizers has been so far poorly addressed.[14,19,61] Only few studies exposed soil organisms to MPs. Nevertheless, the impacts observed are not as high as those reported for other CECs. In a study conducted by Cao et al.,[79] the earthworms

(*Eisenia fetida*) were exposed to polystyrene (PS) MPs (size of 58 μm) in soils at different concentrations (between 0% and 2% w/w). Non-significant effects were observed when earthworms were exposed at concentrations below 0.5% while, for higher concentrations, the presence of PS inhibited their growth and increased the mortality, possibly due to the damage in their self-defense system. Other studies developed by Huerta Lwanga et al.[80] with *Lumbricus terrestris* earthworms revealed some effects when they were exposed to low-density polyethylene, ranging from 200 to 300 μm, at different concentrations (between 0% and 70% in soil). Specifically, the presence of MPs affected the borrowing activity of earthworms[80,81] and this would result in an increase of the sorption of other pollutants as described elsewhere.[82] In a more recent work by the same authors, they observed that the gut bacteria of *Lumbricus terrestris* earthworms was involved in the decay of low-density polyethylene MPs by reducing the particle size.[83] The epigeic earthworms *Eisenia andrei* were also exposed to polyethylene, between 62.5 and 1000 mg MP/kg soil (dw), and sizing between 250 and 1000 μm, and different toxic effects were observed including damages at the histopathological level and immune system response. However, no significant effects were recorded on survival, number of juveniles or final weight of adults.[84] For MPs, some authors have pointed out the necessity to evaluate the impact for soil microbiome in order to capture the emergent properties of a system and not only to assess risks with individual-level trait studies.[85] Indeed, the results of a study that exposed the soil microbiome to mulch residues of polyvinyl chloride film showed that the functional diversity (community level physiological profile) as well as the enzyme activity of dehydrogenase and fluorescein diacetate hydrolysis slightly decreased with the increase of plastic concentration.[86] However, in these studies, the effect of nano and microparticles size has not been evaluated.

For CBENMs, literature on their effects on soil biota is extensive due to their potential beneficial applications in agriculture.[28,87,88] Effects reported are widely diverse. For fullerenes, they range from having almost no impacts in soil microbial communities[89] to suppress from three up to four-fold the growth of fast growing bacteria.[90] However, as it occurs for PPCPs, most of the available studies to evaluate risks in soils are based on the estimation of RQs, using PEC and PNEC values.[35,63,91] In a general extent, RQs for CNTs and fullerenes in sludge-treated soils were far below one.[35] Another strategy to estimate risks is the use of the probabilistic species sensitivity distributions (pSSDs) method, which accounts for the inherent uncertainty and variability of effects to obtain PNEC values.[92]

This approach provides data about how sensitive organisms are when exposed to CBENMS and allows to estimate hazard concentrations (HCx%). For example, HC5 means that 5% of the organisms are affected, while 95% are unaffected. The remaining 95% of non-affected organisms is the lower confidence limit of such an HC value, which is often used and combined with assessment factors to derive a PNEC. Coll et al.[63] used pSSDs to obtain PNECs and calculated risk characterization ratios (RCR) as PEC/PNEC. As it occurs for RQs, the authors considered that no risk would be expected if $RCR < 1$, while $RCR > 1$ would indicate that adverse effects might occur on terrestrial organisms, thus suggesting the need to implement additional risk management measures. In general, RCR obtained were <0.01 in sludge-treated soils for both fullerenes and CNTs,[63] suggesting that these compounds do not pose a significant risk to the terrestrial environment.[35,63]

3.3 Transport in soil and contamination of natural water resources

CECs may also be transported through the soil column, leading to their occurrence in aquifers, streams or drainage channels adjacent to the agricultural fields. Several studies revealed the potential runoff of PPCPs from sludge-amended fields after rainfall events. A recent review from Ghirardini and Verlicchi[93] compiles information about 38 studies performed between 1980 and 2017 on the occurrence of 57 PPCPs, in land runoff and tile drainage from fields fertilized with treated sewage sludge. The most frequently detected compounds were hormones, the antiseptics triclosan and triclocarban, the analgesics and anti-inflammatories acetaminophen, ibuprofen and naproxen, the antibiotic sulfamethoxazole, the lipid regulator gemfibrozil and the psychiatric drug carbamazepine. For surface runoff, PPCPs concentrations were measured in different events after sludge application. For some compounds such as triclosan, atenolol, acetaminophen and sulfamethoxazole, the highest concentrations correspond to the first rainfall event after sludge application, while for others (i.e., naproxen, triclocarban, carbamazepine and caffeine) runoff may occur during subsequent events.[93] Concentrations in runoff ranged from <LOD up to approximately 1500 ng/L, while for tile drainage they varied between 1.5 and 41,117 ng/L. To the best of our knowledge, information on the occurrence of PPCPs in leachates and groundwater is very limited. A study conducted by Bourdat-Deschamps[54] analyzed PPCPs in leachates from soils amended with organic wastes, including sewage sludge, using passive capillary wick samplers. The frequencies of

PPCP detections was very low (<7%) and only few substances, such as carbamazepine, ibuprofen, ofloxacin and sulfamethoxazole, were detected at concentrations generally below 0.1 μg/L. Gottschall and co-workers investigated the occurrence of PPCPs[94] and hormones[70] in groundwater for a period of 1 year after land application of dewatered sewage sludge and these compounds were not detected in groundwater at >2m depth below the soil surface. Positive detections at 2m depth were only observed once, just after the first rain event following sludge application. The compounds detected were ibuprofen, triclosan, triclocarban and venlafaxine at concentrations ranging from 10 to 19ng/L. The limited transport of these compounds to groundwater bodies may be attributed to a combination of the PPCPs physicochemical properties (i.e., hydrophobicity) and the limited macroporosity of the field soils.[70]

For MPs, information about their transport to natural water resources is almost inexistent, mostly attributed to the wide range of microparticles types (i.e., fibers, films, other particles) and sizes available, which difficulties the simulation of their behavior in the soil column.[14] Engdahl et al.[95] modeled the transport of MPs in porous media such as soils, indicating that sometimes counter-intuitive transport could take place for MPs. Theoretically, the mobility of MPs in soils depends on plastic size (i.e., nanoplastics would have more mobility than MPs), hydrophobicity, charge, density and shape (i.e., fiber, bead, planar structure, foam).[96] Furthermore, the mobility of MPs in soils would be affected by other factors, such as the possibility of MPs to be surrounded by an ecocorona of microbes and various organic deposits that would modulate their transport in soils,[96,97] the degradation of biopolymers,[96] soil properties (i.e., texture, pH, cation exchange capacity, and soil organic matter content)[62,98] and agricultural practices.[85]

For CBENMs, available studies suggest that these compounds are more likely to adsorb to, and interact with, clay mineral surfaces[99] and thus, have low mobility along the soil column. Experiments performed in water saturated artificial columns, demonstrated that fullerenes have a high affinity and retention in soil.[100–102] Another study performed in natural soils amended with biosolids, containing spiked amounts of C_{60} fullerene, evidenced that only small fractions of C_{60} (<5% of the applied amount) leached from the biosolid/soil system.[103] Indeed, in this study, observed losses of C_{60} in the mass balance suggested its potential biotransformation to more hydrophilic species such as fullerol (C_{60}-OH), which have more affinity for the aqueous phase, and thus may enhance extractability within the leaching process. For CNTs, despite their hydrophobicity, some studies showed that SWCNTs might significantly increase their mobility in the presence of mesoporous and organic matter.[104]

3.4 Plant/crop uptake

Several studies reported that PPCPs can be taken up by crop plants through roots and translocated to the aerial parts of the plants.[105–110] Their uptake and transport within the plant vascular translocation system (xylem and phloem) largely depend on their physical-chemical properties (i.e., hydrophobicity and ionization behavior) as well as on soil properties (such as pH, organic matter content, soil type, etc.),[72,108,111] the plant's physiology and transpiration rate[111,112] and environmental conditions.[113] Most of the available studies performed up to date have been performed under hydroponic and greenhouse-controlled conditions, using spiked culture medium and soil[114–119] or sludge mixed with soil,[120–123] while those focusing on real agricultural systems, representing actual farming practices, and real environmental conditions, are very limited.[94,124,125]

Hydroponic and controlled experiments using spiked soil showed the ability of a wide variety of crop plants used for human consumption, such as radish, ryegrass, silverbeet, lettuce, cucumber, tomato, collards, grass, watercress, spath, carrot, barley, cabbage or cauliflower, to take up PPCPs of common use. Compounds such as carbamazepine, oxazepam, salbutamol, tetracycline, sulfamethazine, sulfamethoxazole and triclosan showed a relevant accumulation potential in crops.[114–116,118,119] In general, for most PPCPs, root concentration factors were higher than those found in leaves. Nevertheless, concentrations detected in leaves were also relevant, showing their translocation potential.[114,115,118,119,126] Greenhouse controlled experiments using agricultural soils amended with sewage sludge also showed great uptake potential of several PPCPs in crop plants[120–123] such as soybean, pepper, tomato, collard, lettuce, radish, Chinese cabbage and spinach (see Table 4 for further details). The psychiatric drugs carbamazepine and sertraline, the antiseptics triclosan and triclocarban, the anti-histaminic diphenhydramine, salbutamol (used to treat asthma) and the pain killer tramadol were the compounds detected in plant tissues, including roots. As it has been observed for hydroponic experiments, root concentrations factors are higher than those observed for shoots or leaves. Nevertheless, concentrations in these tissues are also relevant for some PPCPs (see Table 4).

For field experiments, no conclusive data has been obtained so far and results are in contrast with the findings observed in greenhouse-controlled experiments (see Table 4). Gottschall et al.[94] performed an experiment where dewatered municipal sludge was applied to soils. The occurrence of 80 PPCPs was monitored in sludge aggregates incorporated into soil

Table 4 Examples of selected controlled experiments and field studies to assess the uptake of PPCPs by crops.

Target compounds	Plant species	Experimental conditions	Soil properties and PPCPs concentrations	Concentrations in plant tissue	References
Screening of 63 PPCPs and metabolites	Spinach (*Spinacia oleracea* L.)	Greenhouse conditions. Seven different soils amended with un-spiked sewage sludge and packed in columns Temperature: 20–24 °C Air humidity: 30–40% Spinach seeds germinated in a horticultural substrate. After 20 days, spinach plants were moved to the soil-sludge columns. Columns irrigated with tap water After harvest (45 days) plants were separated into roots and leaves	Clay content: 5.01–29.95%; silt: 4.51–65.98%; sand: 6.17–90.48%; organic carbon content: 0.41–2.53%; pH: 4.10–7.32	Highest concentrations in leaves: Sertraline: 7.7–200 ng/g d.w. Carbamazepine: 15–180 ng/g d.w. Carbamazepine-10,11-epoxide: 28–440 ng/g d.w. Metoprolol acid: <LOQ-210 ng/g d.w. Tramadol: 6.6–25 ng/g d.w. Highest concentrations in roots: Sertraline: 46–550 ng/g d.w. *N*-desmethylsertraline: <LOQ-210 ng/g d.w. Telmisartan: 83–740 ng/g d.w. Citalopram: 7.1–68 ng/g d.w. Metoprolol acid: 2.4–72 ng/g d.w. Carbamazepine: 5.8–63 ng/g d.w. Tramadol: 2.6–8.3 ng/g d.w.	120
Carbamazepine (CBZ), diphenhydramine (DIP), triclocarban (TCC)	Pepper (*Capsicum annuum*), tomato (*Lycopersicon esculentum*), collard (*Brassica oleracea*), lettuce (*Lactuca sativa*) and radish (*Raphanus sativus*)	Spiked anaerobically digested sewage sludge and mixed with soil at a ratio of 1:7.3 One plant/container. For radish 10 seeds/container planted and thinned to three once spouted Plants grown outdoors and irrigated with groundwater Tomato and pepper harvested 54 days after planting, while collard, radish and lettuce after 29 days	Sandy clay loam soil: 51% sand, 21% silt and 28% clay Soil pH 7.6; organic matter content 3% 100 ng/g	TCC had the highest root concentrations (200–550 μg/kg) followed by CBZ (100–200 μg/kg) and DIP (5–12.5 μg/kg) In the shoot, CBZ had the highest concentration (300–∼1200 μg/kg) followed by TCC (100–400 μg/kg) and DIP (6–15 μg/kg) The concentration of CBZ in the shoot system decreased in the order pepper > collard > lettuce > tomato > radish. CBZ only found in the fruits of pepper. DIP and TCC only detected in the fruits of tomato and pepper	121

Carbamazepine (CBZ), sulfamethoxazole (SMX), salbutamol (SAL), triclosan (TCS), trimethoprim (TMP)	Chinese cabbage (*Brassica campestris*)	Greenhouse experiment (A) Spiked soil (45% soil; 45% sphagnum peat mass and 10% perlite) (B) Biosolids-amended soil (98% soil and 2% sludge)	(A) (41% sand, 46% silt, 13% clay); pH 6.7; organic matter 20.1% 0.5–2.6 ng/g (B) 41% sand, 46% silt, 13% clay; pH 6.5; organic matter 2.4% 24.7–433.7 ng/g	(A) Aerial and roots: CBZ (255.4 and 272.9 ng/g); SMX (222.8 and 260.3 ng/g); SAL (108.3 and 140.6 ng/g) and TMP (20.6 and 53.7 ng/g) (B) Aerials and roots: CBZ (317.6 and 416.2 ng/g); SAL (21.2 and 187.6 ng/g) and TCS (22.9 and 1220.1 μg/kg)	122
Carbamazepine (CBZ), diphenhydramine (DIP), fluoxetine (FLU), triclosan (TCS) Triclocarban (TCC)	Soybean (*Glycine max* (L.) *Merr.)*	Spiked sludge and mixed with soil; field application of 5 dry tones/acre Pots planted with four seeds at 24/21 °C day/night temperatures with 25–50% humidity. Photoperiod was 16 h. Pots watered with modified Hoaglands nutrient solution to pH 6.5 Plants harvested after 60 and 110 days. Plants separated into root, stem, leaf and bean (only after 110 days)	Sandy soil with 9.7% clay, 3.2% silt and 87.1% sand, with pH 5.1 and 2.7% organic matter. Soil spiked at 70 ng/g	After 60 days: CBZ, TCS and TCC accumulated in roots and translocated into above ground parts; CBZ (216 ± 75 ng/g in leaf and 153 ± 46 ng/g in root); TCC (7.1 ± 1.8 ng/g in leaf and 126 ± 31 ng/g in root); TCS (28.9 ± 6.1 ng/g in root and 17.1 ± 3.3 ng/g in leaf). DIP and FLU detected in root at very low levels and translocation to above ground tissues was limited. After 110 days: CBZ highest concentrations in leaves (127 ± 30 ng/g root and 110 ± 25 ng/g leaf); TCC: (168 ± 34 ng/g root and 2.6 ± 0.3 ng/g leaf); TCS: (76.8 ± 3.1 root and 120 ± 37 ng/g in leaf)	123

Continued

Table 4 Examples of selected controlled experiments and field studies to assess the uptake of PPCPs by crops.—cont'd

Target compounds	Plant species	Experimental conditions	Soil properties and PPCPs concentrations	Concentrations in plant tissue	References
Carbamazepine 10,11-epoxide-carbamazepine Dihydroxy-carbamazepine	Tomato (*Solanum lycopersicum*), wheat (*Triticum aestivum*) and Romaine lettuce (*Lactuca sativa*)	Field experiments using lysimeters. Comparison between irrigation with treated wastewater and biosolids application Composted biosolids application from 0 to 60 m^3/ha		Three target compounds detected in all plants Total concentrations in tomato leaves <40 ng/g and fruit <1 ng/g Total concentration in wheat leaves <5–<10 ng/g and ears ND^{-1} ng/g. Total concentration in lettuce leaves: <5 ng/g	124
141 PPCPs and TPs	Tomatoes (variety H9909), potatoes (*Solanum tuberosum*), carrots (*Daucus carota*) and sweet corn (*Zea mays*)	1-year field experiment Anaerobically digested and dewatered biosolids applied to plots at a rate of 8000 dry kg/ha. Experiments in triplicate. Control plots with no biosolids application were used	Silt loam soil (gray brown Luvisol) pH 7.5, sand/silt/clay (%) 18/67/15, organic matter content 3.4%	Two compounds detected in two treated samples: Ciprofloxacin (1.76–2.14 ng/g d.w.) in sweet corn and progesterone (0.32–0.34 ng/g d.w.) in potato Six compounds detected in one treated sample: Glyburide (0.6 ng/g) and naproxen (0.8 ng/g) in sweet corn; triamterene (0.04 ng/g) in carrot; 4-epianhydrotetracycline (6.25 ng/g); caffeine (4.24 ng/g) and atenolol (0.5 ng/g) in potato	125
80 PPCPs	Winter wheat (Pioneer, soft red winter wheat 25R47)	Field experiment with dewatered municipal biosolids applied at a rate of 22 Mg dw ha^{-1} Soil cores (0–03 cm) taken at: (a) pre-application; (b) ~6 months post-application; (c) ~1-year post-application Wheat grain sampled during harvesting after 1-year application	Sand (17–75%), silt (17–43%), clay (6–52%); %OC (0.1–1.41); pH (5.3–7.4) No PPCPs detected in soil in pre-application	No PPCPs detected in wheat grain	94

and in wheat grain grown in the field for a period of 1 year following sludge application. Even though PPCPs were detected in the sludge aggregates, no residues were found in wheat grain. In another field experiment the uptake of 141 PPCPs and transformation products (TPs) was evaluated in tomatoes, carrots, potatoes and sweet corn in soils fertilized with municipal sewage sludge.[125] Only 8 out of the 141 target analytes were detected in 1 or 2 crop replicates at concentrations ranging from 0.33 to 6.25 ng/g dry weight (dw), but no analytes were consistently detected above the detection limit in all the triplicate treated plots monitored, suggesting that the potential for PPCP uptake into crops under normal farming conditions is low. Finally, Mordechay et al.[124] studied the uptake of carbamazepine and two of its TPs (10,11-epoxide-carbamazepine and dihydroxy-carbamazepine) in tomato, wheat and lettuce irrigated with treated wastewater and in sludge-amended soils. Results showed that all three compounds were detected in all the crops grown under all treatments. Nevertheless, bioconcentration factors were significantly lower in the soils amended with biosolids compared to those irrigated with treated wastewater.

Following uptake, PPCPs (especially pharmaceuticals) may undergo biotransformation within the plants,[115,127,128] and in some cases, the amounts of metabolites formed can be even greater than the accumulation of the parent compound.[111,129,130] Metabolic enzymes, such as hydrolases or cytochrome P450 enzymes, transform PPCPs once they are in the plant cells, creating a variety of TPs that are eventually mineralized or incorporated into the plant tissues.[131] Compounds such as analgesics and non-steroidal anti-inflammatory drugs, psychiatric drugs (i.e., sertraline, citalopram, carbamazepine, lamotrigine and benzodiazepines), veterinary drugs (i.e., carbadox, flubendazole), hormones (i.e., 17β-estradiol and estrone), antibiotics (i.e., sulfamethoxazole, sulfadiazine), antibacterial agents (triclosan) and stimulants (i.e., caffeine) showed metabolism in the plants.[115,127,132,133]

In the case of MPs, their uptake by plants is not expected due to their large size and the high molecular weight. However, this would not be the case for nanosized polymers.[85] For example, it was observed that fluorescent nanopolystyrene beads (20–40 nm) could enter tobacco cells via endocytosis,[134] or they could be taken up and accumulated in edible plants, such as lettuce, posing a potential risk for humans.[135] For MPs the direct plant uptake cannot be considered as the only environmental problem. Other indirect effects that MPs and nanosized plastics could have are related with the availability of plant nutrients (i.e., films in soils create channels for water movement which could lead to an increase of water evaporation and, consequently soil drying with negative effects for plants[136]); non-degradable polymers may immobilize nutrients or they could alter the C:N soil ratio when these polymers are biodegradable[137]; the transport or adsorption of other co-contaminants by MPs which would have been less available to soil biota and plants in other conditions[137]; and changes in soil microbial community that could alter the soil composition and, thus, plants and crops.[85]

For CBENMs, and specially CNTs, there are numerous studies that report their uptake by various crop species[87,138] (see Table 5 for detailed examples). Due to analytical limitations, the information about concentration levels of these compounds in plant tissues is scarce.[138] For this reason, the use of radio labeled CBENMs to quantitatively determine their uptake by plants has emerged as the most promising technique. For instance, ^{14}C-labeled C_{60} was used to evaluate the sorption of fullerenes in soils and their plant uptake in radish.[140] The authors found that ~7% of the initial fullerenes dose was taken up by plants, with half of this amount accumulating in roots (40–47%), and the remaining translocated in the tuber (22–23%), stem (12–16%) and leaves (18–22%) (Table 5). Another study also using radiolabeled CBENMs evaluated the bioaccumulation of fullerol (^{13}C-C_{60}-OH) in wheat (*Triticum aestivum* L.).[86] Bioaccumulation in the plant, calculated as mass percentage of dose per gram of tissue, ranged from 85.7% to 263.9% initial dose (ID) per g, depending on the dose and growth time. As in the previous study, most of the particles accumulated in the roots with minor translocation occurring to stems and leaves (<4.13% ID per g). Gogos et al.[142] quantified MWCNTs in wheat (*Triticum* spp.) and corn (*Z. mays*), with ~9.8% of the initial amount applied into the soil being translocated to the shoots in a treatment performed at 3 mg/kg, while only 0.015% of the initial amount translocated at 2933 mg/kg.

3.5 Phytotoxic effects

Uptake of CECs might induce some undesired toxic effects in crops. Several publications already highlighted examples of phytotoxicity in plant species when exposed to PPCPs.[143,144] A large proportion of these studies are focused on antibiotics, particularly tetracyclines and sulfonamides,[145–148] psychiatric drugs (carbamazepine and venlafaxine)[127,149,150] and analgesics and anti-inflammatories,[127,151] with little research demonstrating toxic effects resulting from the exposure to other classes of PPCPs.[149] The most studied phytotoxic endpoints are germination, growth inhibition, plant development, tissues deformation, reduced photosynthetic rate and chlorophyll content, disturbances in redox homeostasis and other stress-related

Table 5 Examples of experiments for CBENMs uptake in plants.

Target compounds	Plant species	Experimental conditions	CBENMs concentrations and soil properties	Concentrations in plant tissue/ distribution in plant tissue	References
Fullerol (^{13}C-C_{60}-OH) [$C_{60}(OH)_{20}$]	Bitter melon (*Momordica charantia*)	– Seed pre-treatment: • in fullerol solutions for 48 h • seeds planted in soil pots – Plant culture: • Pot experiments kept in greenhouse lab scale • Temperature (80/65 °F at day/night), relative humidity (70%) and photoperiod (16/8 h light/dark) • Pots watered once in a day • Peter Excel (Everris) fertilizer solution of 15:15:15 of N, P and K, respectively applied in the pots once in a week	– 0–10 μg/mL – 3B potting mix (Fafard)	– Uptake in the roots – Translocation to flowers and fruits – Bioaccumulation in tissues and cells of root, stem, petiole, leaf, flower and fruit at certain concentrations	139

Continued

Table 5 Examples of experiments for CBENMs uptake in plants.—cont'd

Target compounds	Plant species	Experimental conditions	CBENMs concentrations and soil properties	Concentrations in plant tissue/ distribution in plant tissue	References
Fullerol (^{13}C-C_{60}-OH) [$C_{60}(OH)_{20}$]	Wheat (*Triticum aestivum* L.)	– Seed pre-treatment: • surface-sterilized in a 15% NaClO solution for 20 min and washed three times with deionized water • rinsed thoroughly with running deionized water for 24 h • Seeds placed in 10 cm × 1.5 cm Petri dishes with filter paper for germination • 1 week for germination – Plant culture: • Lab scale growth chamber incubation at different growth times • in a dark incubator at 25 °C • alternation of day and night (12 h:12 h) at light strength of 24,000 lx for the day cycle • After 1 week, sets of three seedlings of uniform size transferred to 100 mL beakers	– 2.5–10 μg/mL	– Uptake of 85.68–263.86% ID per g, (mass percentage of dose per gram tissue) in roots – Translocation from roots to stems and leaves of <4.13% ID per g – Bioaccumulation: • increasing concentrations mainly in the roots and much lower in stems and leaves • ~7% ED/g (percentage of exposed dose per gram) • significant fullerenol concentration and culture time dependence: o long-term and 2.5 μg mL^{-1} → significant content increase in roots o long-term and 10.0 μg/mL → accumulation suppression	86

		• Each beaker fertilized with 50 mL of Hoagland improved nutrient solution, which contained (all in mgL^{-1}) $Ca(NO_3)_2{\cdot}4H_2O$ (945), $KNO_3{\cdot}4H_2O$ (506), NH_4NO_3 (80), KH_2PO_4 (136), and $MgSO_4$ (493) and the micronutrients $FeSO_4{\cdot}7H_2O$ (13.9), EDTA·2Na (18.65), KI (0.83), H_3BO_3 (6.2), $MnSO_4$ (22.3), $ZnSO_4{\cdot}7H_2O$ (8.6), $Na_2MoO_4{\cdot}2H_2O$ (0.25), $CuSO_4{\cdot}5H_2O$ (0.025), and $CoCl_2{\cdot}6H_2O$ (0.025). • nutrient solution added daily to ensure constant total volume at 50 mL • 7 days of exposure			
C_{60} fullerene	radish (*Raphanus sativus*)	– Seed pre-treatment: • germinated in a Petri dish and then transferred to 250-mL beakers containing 100 g of sand and in containers using potting soil – Plant culture: • Hydroponic experiments in an environmental chamber • light (14 h:10 h light:dark) and a temperature of 25 °C • at 4 leaf stage, dosed with 5 mL solutions of ^{14}C-labeled C_{60} in water • 13–16 days exposure	– 8.44 mg/L – Sand	– Uptake in roots of ~7% – Translocation from roots to tuber, stem and leaves – Bioaccumulation mainly in roots (40–47%); and smaller amounts in tuber (22–23%), stem (12–16%), and leaves (18–22%)	140

Continued

Table 5 Examples of experiments for CBENMs uptake in plants.—cont'd

Target compounds	Plant species	Experimental conditions	CBENMs concentrations and soil properties	Concentrations in plant tissue/ distribution in plant tissue	References
MWCNTs	*A. thaliana*, rice (*Oryza sativa* L.), maize, and soybean (*Glycine* max)	– Seed pre-treatment: • sterilized with ethanol (70%, *v*/v) 10 min and twice soaked in Clorox (25%, *v*/v) 15 min • washed five times with autoclaved deionized water before use. • Twenty-five sterilized seeds were placed on Petri dishes • stratified at 4 °C for 24 h • transferred to the growth chamber – Plant culture: • Hydroponic experiments in environmental chamber • light period: 16/8 h; temperature: 22/20 °C, day/night • 14 days incubation • transferred to Hoagland's solution in 250 mL beaker • 5–7 days acclimatization • moved into 90 mL fresh Hoagland's solution amended with 2.25 mg/L MWCNT suspension • 1-day exposure	– 2.25 mg/L	– Uptake in roots at different percentages according to plant species – Translocation from roots to stems and leaves → Calculated the translocation factor (TF: ratio of average MWCNTs content in leaves and stems vs in roots): • TF = 0.582 in rice > 0.262 in Arabidopsis > 0.054 in maize > 0.141 in soybean – Bioaccumulation ranging from 0.53 ± 0.31 (in maize sheath) to 76.6 ± 6.1 (in soybean root) mg/kg • accumulation of MWCNTs in *A. thaliana* and soybean ~1.5- to 3-fold > rice and maize • MWCNTs content in rice: • root = 22.5 mg/kg • ≈2-fold sheath; ≈ 10-fold leaf • maize and soybean accumulation in roots ≫ aboveground tissues, MWCNT content in Arabidopsis leaf (13.0 ± 0.9 mg/kg) was the highest • MWCNTs content in soybean root ≈ 1.4-fold Arabidopsis; ≈ 3.4-fold rice; ≈2.4-fold maize	141

MWCNTs	Red clover	– Plant culture: • Greenhouse pot experiment at lab scale • Homogenously spiked natural agricultural soil • seven pot replicates consisting of seven plants per pot • pot filled with a drainage layer of sand (0.5 L, 520 g) and 3.3 kg soil (corresponding to 2.9 L). • pot was kept at 60–70% of the total water holding capacity (WHC) during the entire experiment. • 16 h light period (light intensity of 300 W m^{-2}) and a 25/16 °C light/dark temperature regime • plants fertilized after 9 weeks • 14 weeks exposure	– 3 and 2933 mg/kg – Soil: • sandy loamy to loamy fine fraction • Properties of spiked soil-quartz mixture (50:50 v/v): 0.55 (±0.03) % Org. C, pH 7.7, 86.1% sand, 6.3% silt, 6.7 (±0.5) % clay	– Uptake in plant: • 68 (3 mg/kg treatment, $n=1$) and 99 µg/g (2933 mg/kg treatment, $n=1$). • 0.97 and 1.5 mg of MWCNTs taken up into the plants per pot in the two treatments, respectively • 9.8% and 0.015% of the initial MWCNT amount in the soil taken up in 3 and 2933 mg/kg treatments, $n=1$, respectively	142

phenomena.[107,152–155] Nevertheless, most of these studies were based on in vitro assays, rather than under real soil conditions, using much higher exposure concentrations for individual substances than those observed in real agricultural environments.[143]

For MPs, effects on the germination rate of cress (*Lepidium sativum*) were observed, which were significantly reduced when exposed to green fluorescent plastic particles between 50 and 4800 nm.[156] Similar results were observed by Qi et al.[157] when wheat seeds (*Triticum aestivum*) were exposed to 1% of polyethylene and biodegradable polyethylene terephthalate in soil. Both plastic types affected wheat growth, and biodegradable MP had more negative effects than the non-degradable polymer. Interestingly, the authors also observed that the negative effects for wheat seeds were attenuated by the presence of soil earthworms.[157]

The effects of CBENMs are broad and diverse, from overt phytotoxicity to positive effects, such as the promotion of plant growth and/or productivity to control diseases.[138,158–160] This heterogeneity of effects depends on CBENMs properties and doses applied, experimental conditions and plant/microorganism type.[158] After uptake and accumulation in plants, CBENMs can affect factors involved in plant growth such as seed germination rate, root length and plant weight, and be responsible for changes in processes such as photosynthesis, nitrogen fixation, transpiration rate, lipid peroxidation and reduction in stress-related genes.[160] For instance, exposure to low concentration (100 μg/mL) of MWCNT in soil promoted nitrogen fixation activity of nodules in a legume-Rhizobium (*Lotus japonicus* and *Mesorhizobium loti*) system and consequently increased plant biomass by 14–25%.[161] On the other hand, zucchini (*Cucurbita pepo* L.) seeds exposed to MWCNTs decreased seed germination, root length, bud length, biomass accumulation and vitality index.[162] Numerous studies have demonstrated the potential of CBENMs to exert beneficial effects in plants, and this has prompted their application in plant science/agriculture, aiming at enhancing crop growth and production.[88] For instance, bitter melon seeds (*Momordica charantia*) were exposed to different concentrations of fullerol [$C_{60}(OH)_{20}$], to evaluate its effects on the plant biomass yield and fruit characters,[139] resulting in an increase of up to 54% in biomass, 24% in water content, and 128% in fruit yield.

3.6 Public health implications

Uptake into plants, especially edible crops, may represent an important exposure pathway of CECs into the food chain and thus, present a risk to humans.[118] Several risk assessment studies have been done, most of them focusing on PPCPs. One approach used is the calculation of acceptable daily intake values (ADI), which estimates the amount of a substance that can be consumed by a human without resulting in an appreciable risk to health. This strategy was used by Carter et al.,[118] who showed that if ryegrass and radish were grown in soils containing carbamazepine, diclofenac, fluoxetine, propranolol, sulfamethazine and triclosan, humans would not consume greater levels than ADI for any of the compounds. Kodešová et al.[120] also estimated ADI for PPCPs detected in spinach tissues. Besides ADIs, they also estimated the daily consumption (DC) of fresh spinach leaves by a child (25 kg) and an adult (70 kg) to reach ADI and a similar analysis was done for roots. Results showed that for most compounds, DC values for adults mostly exceeded 1 kg of spinach leaves per day, except for sertraline and 10,11-epoxycarbamazepine, whose DC varied from 0.4–0.5 to 0.76–0.79 kg, respectively, in certain soils. In the case of children, DC values were 2.8 times lower, ranging from 0.1 to 0.2 kg for sertraline and 0.28–0.29 for 10,11-epoxycarbamazepine. The DC threshold values for sertraline in roots were even lower than those observed in leaves (DC for children ranged from 0.03 to 0.42 kg). These results indicate that some risks could occur to human health in the case of a consumption of vegetables planted in soil-sludge substrates. However, the study conducted by Roccaro et al.,[163] showed controversial results, highlighting that no health risks are to be expected. This is confirmed by Prosser et al.,[164] who performed a risk assessment study of the PPCPs found in edible parts of crops grown in biosolids-amended fields, among other fertilization practices. Estimated daily intake values were compared to ADIs and amendment with biosolids or animal manure resulted in hazard quotients ≥ 0.1 for few PPCPs, namely carbamazepine, diphenhydramine, salbutamol, triclosan and sulfamethazine. However, in general, minimum risks to human health are expected.

3.7 Dissemination of antibiotic resistance

AMR is an increasingly serious threat to global public health recognized by the World Health Organization (WHO), who already developed a Global Action Plan on AMR.[165] Although this plan considers the dissemination of AMR in the environment and not only in clinical settings, it does not provide any clear advice about the wastewater treatment or sewage sludge disposal. Therefore, in 2017, the European Commission (EC) adopted the EU One Health Action Plan to face the dissemination of AMR. This plan supports the research into knowledge gaps on the release and spread of ARB and antimicrobials into the environment and the development of technologies for their reduction in the environment.[166]

The fertilization using sewage sludge or biosolids could promote the horizontal gene transfer (HGT) of MGE (i.e., plasmids, integrons, transposons) allowing the interaction between ARB and ARGs with indigenous soil bacteria in the presence of chemical stressors such as PPCPs or heavy metals. It is expected that repeated amendment of agricultural soils with sludge or biosolids might contribute to the dissemination of AMR.[52] Indeed, one study performed in hydroponic systems showed that plants growing in the presence of antibiotics exhibited higher rates of endophytic ARB and ARGs.[167] However, the studies performed so far in soil reported divergent results as they varied due to the differences between culture-dependent and culture-independent methods, between microcosm or long-term field experiments and also due to the nature of the sewage and the treatment of sludge to be converted into biosolid.[52,168,169]

The impact of fertilizing soils with raw sewage sludge was assessed in an experiment, which reported that changes on ARB prevalence were transient and, after 7 months of the amendment, the levels of ARB returned to the initial levels.[170] Another longer field-study analyzed the impact of the application of anaerobically digested biosolids to soils and to the crops.[171] As a result, the authors could not establish any coherent impact on the frequency of AMR in viable coliform bacteria from soil and vegetable samples at harvest. Accordingly, it seems that a delay between sludge application and crop harvest should be enough to decrease the direct risk of transmission of ARB from produce to humans.[171] However, both studies showed that ARB increased in the beginning of the experiments, which could lead to their dissemination to groundwater as it was previously demonstrated how bacteria from land can contaminate groundwater after rainfall events.[69]

Regarding ARGs, many ARGs or genes associated to MGE were detected in soil and vegetables harvested from crops amended with biosolids.[170,171] This is in accordance with a microcosm study where soil was supplemented with mesophilic anaerobic digested biosolids for 6 months. In that case, the ratio between some ARGs (including also class 1 integron *intI1* gene) and 16S rRNA remained constant during the treatment.[172] Integrons can transfer ARGs among microorganisms through typical gene capture and, particularly the *intI1* gene is usually correlated with ARGs.[6,43] Therefore, their persistence in the amended soil suggests that the sludge could promote the maintenance and the HGT of ARGs.

This persistence also increases the time for potential transport of these genes to groundwaters since rainfall events can also increase the concentrations of ARGs in runoff.[173]

3.8 Strategies to reduce environmental risks

As previously mentioned, AD and composting are two of the most widely used technologies in Europe for the biological stabilization of sewage sludge prior to its application on agricultural land.[174,175] However, data available on the literature show a great variability in the efficiency of these technologies for the removal of CECs, ARBs and ARGs, not only associated with the limited biotransformation of some compounds under the redox conditions corresponding to each system but also to the influence of operational conditions applied during sewage sludge treatment.[176–178] In view of these facts, the evaluation of potential strategies to improve conventional technologies by means of the combination with pre-/post-treatment systems and/or the application of alternative treatment technologies is considered essential in order to increase the removal of several CECs and to reduce environmental and human health risks.

The application of pre-treatment technologies for sewage sludge before AD has gained importance in recent years with the aim of accelerating the sludge hydrolysis and increasing subsequent biogas production.[179] According to the literature, the application of pre-treatments combined with AD generally provides extra benefits for ARGs reduction during sludge treatment compared to the use of single AD. However, the ARGs removal efficiency depends on the pre-treatment technology applied.[177,180,181] For instance, Wang et al.[182] observed that the ARGs removal efficiency of the AD process during sewage sludge treatment (~51%) could be improved up to 53–75% including alkaline, thermal hydrolysis and ultrasound pre-treatments. Additionally, the application of pre-treatment technologies, such as thermal pyrolysis, microwave-assisted pyrolysis and catalytic pyrolysis, before AD is also considered a promising option to reduce the concentration of MPs from sewage sludge.[183] On the other hand, recent studies found in the literature propose the application of post-treatment systems after AD as an effective solution to produce digested sludge for a safe agricultural reuse. In this sense, Min et al.[184] and Tomei et al.[185] demonstrated the benefits of post-aerobic treatment after AD in terms of PPCPs and ARGs removal, meanwhile other authors[186,187] consider that post-composting could be one of the best follow up strategies for the additional removal of antibiotics and ARGs from digestate. Among alternative sewage sludge treatment technologies, hyperthermophilic composting and biodrying have emerged recently as an improved option of the conventional composting because both technologies allow to produce natural soil amendments in a shorter period.[188,189] The high temperatures reached during the hyperthermophilic composting (about 90 °C) increase the efficiency of the process to remove some CECs compared to conventional composting. For instance, Chen et al.[190] observed a removal rate for MPs from sewage sludge around 43.7% during the hyperthermophilic composting, which was about nine times higher than the reduction rates observed in conventional composting, whereas Liao et al.[51] demonstrated that hyperthermophilic composting enhances the removal of ARGs and MGEs in sewage sludge due to a higher reduction in bacterial abundances and diversity of potential ARG hosts under hyperthermophilic conditions. In the case of biodrying, different studies have demonstrated its potential to reduce the concentration of PPCPs frequently detected in sewage sludge, as well as ARGs. In this sense, Pilnáček et al.[191] observed that the antibiotic activity was completely reduced during sludge biodrying under mesophilic conditions, whereas Zhang et al.[192] demonstrated its higher efficiency to reduce some ARGs over other sludge treatment technologies, such as AD, air-drying and aerobic digestion, observing lower half-lives for typical ARGs such as *tetA, ermB, tetW* and *sulI.* Another technology that is also receiving an increasing attention as a sustainable process for the conversion of sewage sludge into hydrochar is hydrothermal carbonization.[193] Several authors[194,195] demonstrated the potential of this technology to efficiently reduce the concentration of pharmaceuticals in sewage sludge (e.g., diclofenac and carbamazepine were removed up to 56% and 97%, respectively). Unfortunately, despite thermal conversion of sewage sludge likely affects also the fate of personal care products, ARBs and ARGs, there is still no literature available on this topic.[196,197]

4. Conclusions

Information available up to date proves that CECs are spread into the environment following sludge application as fertilizer in soils. Studies revealed that the target CECs, specially MPs and CBENMs, are persistent in the terrestrial environment. For AMR, the number of studies performed so far suggest that sewage sludge fertilization could be a relevant route of dissemination to the environment. For all CECs, experimental data about the undesired effects on soil microorganisms, functioning, and fauna is still much needed. This is especially relevant for MPs, since information about their fate and ecotoxicological effects in soils is almost inexistent. Current literature also reveals that CECs, especially PPCPs, can be transported from soil to natural and groundwater resources following rainfall events. This is not the case for CBENMs, which are more prone to attach to solid matrices, and thus, their occurrence in water resources is not expected. ARB and ARGs can spread to groundwater interacting with indigenous environmental bacteria and exchanging MGEs by HGT, thus disseminating the AMR to the environment. For MPs, information about their potential transport to water resources is still necessary. Studies performed also indicate that the uptake of target CECs, their translocation and accumulation in edible parts of crops grown in sludge-amended soils in real agricultural systems is feasible. Available investigations also point out that sewage sludge fertilization may represent a direct risk of exposure to ARGs for the consumers. Nonetheless, plenty of knowledge gaps still exist, requiring further research on this topic, especially for MPs, ARBs and ARGs. More studies performed under real agricultural systems are needed, including a wider spectrum of plant species. This data is necessary since it would allow to have more accurate estimations about the potential public health risks associated with the consumption of crops grown under sludge-amended soils. The phytotoxic effects of CECs also warrant further investigation, specially the evaluation of chronic effects. The upgrade of current treatment technologies or the development of alternative sludge treatments stands as a suitable option to minimize potential environmental and human health risks. Nevertheless, the performance of these technologies needs further evaluation.

Acknowledgments

M.L. acknowledges the project DEGBIOPLAST (RTI2018-097860-J-I00) funded by the Spanish Ministry of Economy and Competitiveness. V.O. acknowledges the project PECT-Osona-Transformació Social funded by Fundació Universitària Balmes-PO FEDER, (PRE3084/2018). M.G. acknowledges Generalitat de Catalunya through the Consolidated Research Group (SGR ICRA-ENV 2017 1124) and the funding from CERCA program.

References

1. Burgmann H, Frigon D, Gaze WH, Manaia CM, Pruden A, Singer AC, et al. Water and sanitation: an essential battlefront in the war on antimicrobial resistance. *FEMS Microbiol Ecol* 2018;**94**:fiy101.
2. Collivignarelli MC, Abbà A, Frattarola A, Carnevale Miino M, Padovani S, Katsoyiannis I, et al. Legislation for the reuse of biosolids on agricultural land in Europe: overview. *Sustainability* 2019;**11**:6015–37.

3. EIP-AGRI. *Focus group nutrient recycling. Final report.* https://eceuropaeu/eip/agriculture/en/publications/eip-agri-focus-group-nutrient-recycling-final; 2017.
4. Eid EM, Alamri SAM, Shaltout KH, Galal TM, Ahmed MT, Brima EI, et al. A sustainable food security approach: controlled land application of sewage sludge recirculates nutrients to agricultural soils and enhances crop productivity. *Food Energy Secur* 2020;**9**:e197.
5. Ivanová L, Mackuľak T, Grabic R, Golovko O, Koba O, Staňová AV, et al. Pharmaceuticals and illicit drugs—a new threat to the application of sewage sludge in agriculture. *Sci Total Environ* 2018;**634**:606–15.
6. Ma J, Cui Y, Li A, Zhang W, Liang J, Wang S, et al. Evaluation of the fate of nutrients, antibiotics, and antibiotic resistance genes in sludge treatment wetlands. *Sci Total Environ* 2020;**712**:136370.
7. Murray R, Tien YC, Scott A, Topp E. The impact of municipal sewage sludge stabilization processes on the abundance, field persistence, and transmission of antibiotic resistant bacteria and antibiotic resistance genes to vegetables at harvest. *Sci Total Environ* 2019;**651**:1680–7.
8. Smith SR, Lang NL, Cheung KH, Spanoudaki K. Factors controlling pathogen destruction during anaerobic digestion of biowastes. *Waste Manag* 2005;**25**:417–25.
9. Yang Y, Sik Ok Y, Kim K, Kwon EE, Tsang YF. Occurrences and removal of pharmaceuticals and personal care products (PPCPs) in drinking water and water/sewage treatment plants: a review. *Sci Total Environ* 2017;**596–597**:303–20.
10. Verlicchi P, Zambello E. Pharmaceuticals and personal care products in untreated and treated sewage sludge: occurrence and environmental risk in the case of application on soil—a critical review. *Sci Total Environ* 2015;**538**:750–67.
11. Carr SA, Liu J, Tesoro AG. Transport and fate of microplastic particles in wastewater treatment plants. *Water Res* 2016;**91**:174–82.
12. Mason SA, Garneau D, Sutton R, Chu Y, Ehmann K, Barnes J, et al. Microplastic pollution is widely detected in US municipal wastewater treatment plant effluent. *Environ Pollut* 2016;**218**:1045–54.
13. De Souza Machado AA, Kloas W, Zarfl C, Hempel S, Rillig MC. Microplastics as an emerging threat to terrestrial ecosystems. *Glob Chang Biol* 2018;**24**:1405–16.
14. Corradini F, Meza P, Eguiluz R, Casado F, Huerta-Lwanga E, Geissen V. Evidence of microplastic accumulation in agricultural soils from sewage sludge disposal. *Sci Total Environ* 2019;**671**:411–20.
15. Dris R, Gasperi J, Rocher V, Saad M, Renault N, Tassin B. Microplastic contamination in an urban area: a case study in Greater Paris. *Environ Chem* 2015;**12**:592–9.
16. Li X, Chen L, Mei Q, Dong B, Dai X, Ding G, et al. Microplastics in sewage sludge from the wastewater treatment plants in China. *Water Res* 2018;**142**:75–85.
17. Lassen C, Hansen SF, Magnusson K, Hartmann NB, Jensen PR, Nielsen TG, et al. *Microplastics: Occurrence, effects and sources of releases to the environment in Denmark.* Report no. 8793352808. Danish Environmental Protection Agency; 2015.
18. Talvitie J, Mikola A, Setälä O, Heinonen M, Koistinen A. How well is microlitter purified from wastewater?–A detailed study on the stepwise removal of microlitter in a tertiary level wastewater treatment plant. *Water Res* 2017;**109**:164–72.
19. Zubris KAV, Richards BK. Synthetic fibers as an indicator of land application of sludge. *Environ Pollut* 2005;**138**:201–11.
20. Lusher A, Hurley R, Vogelsang C, Nizzetto L, Olsen M. *Mapping microplastics in sludge.* Norwegian Institute for Water Research; 2017. NIVA Reports.
21. Li Q, Wu J, Zhao X, Gu X, Ji R. Separation and identification of microplastics from soil and sewage sludge. *Environ Pollut* 2019;**254**:113076.
22. Mintenig SM, Int-Veen I, Löder MGJ, Primpke S, Gerdts G. Identification of microplastic in effluents of waste water treatment plants using focal plane array-based micro-Fourier-transform infrared imaging. *Water Res* 2017;**108**:365–72.

23. Magnusson K, Norén F. *Screening of microplastic particles in and down-stream a wastewater treatment plant.* 2014.
24. Brandsma S, Nijssen P, Van Velzen M, Leslie H. *Microplastics in river suspended particulate matter and sewage treatment plants.* Report R14/02. Institute for Environmental Studies. VU University of Amsterdam; 2013.
25. Mahon AM, O'Connell B, Healy MG, O'Connor I, Officer R, Nash R, et al. Microplastics in sewage sludge: effects of treatment. *Environ Sci Technol* 2017; **51**:810–8.
26. Zhang GS, Liu YF. The distribution of microplastics in soil aggregate fractions in southwestern China. *Sci Total Environ* 2018;**642**:12–20.
27. Khot LR, Sankaran S, Maja JM, Ehsani R, Schuster EW. Applications of nanomaterials in agricultural production and crop protection: a review. *Crop Prot* 2012;**35**:64–70.
28. Reddy Pullagurala VL, Adisa IO, Rawat S, White JC, Zuverza-Mena N, Hernandez-Viezcas JA, et al. Fate of engineered nanomaterials in agroenvironments and impacts on agroecosystems. In: Marmiroli N, White JC, Song J, editors. *Exposure to engineered nanomaterials in the environment.* Amsterdam: Elsevier; 2019. p. 105–42.
29. Patil SS, Lekhak UM. Toxic effects of engineered carbon nanoparticles on environment. In: *Carbon nanomaterials for agri-food and environmental applications.* Elsevier; 2020. p. 237–60.
30. Piccinno F, Gottschalk F, Seeger S, Nowack B. Industrial production quantities and uses of ten engineered nanomaterials in Europe and the world. *J Nanopart Res* 2012;**14**:1109–20.
31. Kiser MA, Ladner DA, Hristovski KD, Westerhoff PK. Nanomaterial transformation and association with fresh and freeze-dried wastewater activated sludge: implications for testing protocol and environmental fate. *Environ Sci Technol* 2012;**46**:7046–54.
32. Wang Y, Westerhoff P, Hristovski KD. Fate and biological effects of silver, titanium dioxide, and C60 (fullerene) nanomaterials during simulated wastewater treatment processes. *J Hazard Mater* 2012;**201–202**:7.
33. Yang Y, Wang Y, Hristovski K, Westerhoff P. Simultaneous removal of nanosilver and fullerene in sequencing batch reactors for biological wastewater treatment. *Chemosphere* 2015;**125**:115–22.
34. Holden PA, Klaessig F, Turco RF, Priester JH, Rico CM, Avila-Arias H, et al. Evaluation of exposure concentrations used in assessing manufactured nanomaterial environmental hazards: are they relevant? *Environ Sci Technol* 2014;**48**:10541–51.
35. Gottschalk F, Sonderer T, Scholz RW, Nowack B. Modeled environmental concentrations of engineered nanomaterials (TiO(2), ZnO, Ag, CNT, Fullerenes) for different regions. *Environ Sci Technol* 2009;**43**:9216–22.
36. Sun TY, Gottschalk F, Hungerbuhler K, Nowack B. Comprehensive probabilistic modelling of environmental emissions of engineered nanomaterials. *Environ Pollut* 2014;**185**:69–76.
37. Sun TY, Bornhoft NA, Hungerbuhler K, Nowack B. Dynamic probabilistic modeling of environmental emissions of engineered nanomaterials. *Environ Sci Technol* 2016;**50**:4701–11.
38. Straub TM, Pepper IL, Gerba CP. Hazards from pathogenic microorganisms in land-disposed sewage sludge. *Rev Environ Contam Toxicol* 1993;**132**:55–91.
39. Fijalkowski K, Rorat A, Grobelak A, Kacprzak MJ. The presence of contaminations in sewage sludge—the current situation. *J Environ Manage* 2017;**203**:1126–36.
40. Yang Y, Li B, Zou S, Fang HH, Zhang T. Fate of antibiotic resistance genes in sewage treatment plant revealed by metagenomic approach. *Water Res* 2014;**62**:97–106.
41. Ju F, Beck K, Yin X, Maccagnan A, McArdell CS, Singer HP, et al. Wastewater treatment plant resistomes are shaped by bacterial composition, genetic exchange, and upregulated expression in the effluent microbiomes. *ISME J* 2019;**13**:346–60.

42. Hrenovic J, Ivankovic T, Ivekovic D, Repec S, Stipanicev D, Ganjto M. The fate of carbapenem-resistant bacteria in a wastewater treatment plant. *Water Res* 2017; **126**:232–9.
43. Bengtsson-Palme J, Hammaren R, Pal C, Ostman M, Bjorlenius B, Flach CF, et al. Elucidating selection processes for antibiotic resistance in sewage treatment plants using metagenomics. *Sci Total Environ* 2016;**572**:697–712.
44. Mao D, Yu S, Rysz M, Luo Y, Yang F, Li F, et al. Prevalence and proliferation of antibiotic resistance genes in two municipal wastewater treatment plants. *Water Res* 2015;**85**:458–66.
45. He P, Zhou Y, Shao L, Huang J, Yang Z, Lu F. The discrepant mobility of antibiotic resistant genes: evidence from their spatial distribution in sewage sludge flocs. *Sci Total Environ* 2019;**697**:134176.
46. Liu H, Zhou X, Huang H, Zhang J. Prevalence of antibiotic resistance genes and their association with antibiotics in a wastewater treatment plant: process distribution and analysis. *Water* 2019;**11**:2495.
47. Calero-Caceres W, Melgarejo A, Colomer-Lluch M, Stoll C, Lucena F, Jofre J, et al. Sludge as a potential important source of antibiotic resistance genes in both the bacterial and bacteriophage fractions. *Environ Sci Technol* 2014;**48**:7602–11.
48. Munir M, Wong K, Xagoraraki I. Release of antibiotic resistant bacteria and genes in the effluent and biosolids of five wastewater utilities in Michigan. *Water Res* 2011; **45**:681–93.
49. Tian Z, Zhang Y, Yu B, Yang M. Changes of resistome, mobilome and potential hosts of antibiotic resistance genes during the transformation of anaerobic digestion from mesophilic to thermophilic. *Water Res* 2016;**98**:261–9.
50. Su JQ, Wei B, Ou-Yang WY, Huang FY, Zhao Y, Xu HJ, et al. Antibiotic resistome and its association with bacterial communities during sewage sludge composting. *Environ Sci Technol* 2015;**49**:7356–63.
51. Liao H, Lu X, Rensing C, Friman VP, Geisen S, Chen Z, et al. Hyperthermophilic composting accelerates the removal of antibiotic resistance genes and mobile genetic elements in sewage sludge. *Environ Sci Technol* 2018;**52**:266–76.
52. Wolters B, Fornefeld E, Jechalke S, Su JQ, Zhu YG, Sorensen SJ, et al. Soil amendment with sewage sludge affects soil prokaryotic community composition, mobilome and resistome. *FEMS Microbiol Ecol* 2019;**95**:fiy193.
53. Walters E, McClellan K, Halden RU. Occurrence and loss over three years of 72 pharmaceuticals and personal care products from biosolids—soil mixtures in outdoor mesocosms. *Water Res* 2010;**44**:6011–20.
54. Bourdat-Deschamps M, Ferhi S, Bernet N, Feder F, Crouzet O, Patureau D, et al. Fate and impacts of pharmaceuticals and personal care products after repeated applications of organic waste products in long-term field experiments. *Sci Total Environ* 2017; **607–608**:9.
55. Kinney CA, Furlong ET, Kolpin DW, Burkhardt MR, Zaugg SD, Werner SL, et al. Bioaccumulation of pharmaceuticals and other anthropogenic waste indicators in earthworms from agricultural soil amended with biosolid or swine manure. *Environ Sci Tech* 2008;**42**:1863–70.
56. Golet EM, Strehler A, Alder AC, Giger W. Determination of fluoroquinolone antibacterial agents in sewage sludge and sludge-treated soil using accelerated solvent extraction followed by solid-phase extraction. *Anal Chem* 2002;**74**:5455–62.
57. Golet EM, Xifra I, Siegrist H, Alder AC, Giger W. Environmental exposure assessment of fluoroquinolone antibacterial agents from sewage to soil. *Environ Sci Tech* 2003;**37**:3243–9.
58. Pulkrabová J, Černý J, Száková J, Švarcová A, Gramblička T, Hajšlová J, et al. Is the long-term application of sewage sludge turning soil into a sink for organic pollutants? evidence from field studies in the Czech Republic. *J Soil Sediment* 2019;**19**:2445–58.

59. Albero B, Sánchez-Brunete C, Miguel E, Pérez RA, Tadeo JL. Analysis of natural-occurring and synthetic sexual hormones in sludge-amended soils by matrix solid-phase dispersion and isotope dilution gas chromatography-tandem mass spectrometry. *J Chromatogr A* 2013;**1283**:39–45.
60. Butler E, Whelan MJ, Ritz K, Sakrabani R, Van Egmond R. Effects of triclosan on soil microbial respiration. *Environ Toxicol Chem* 2011;**30**:360–6.
61. Nizzetto L, Futter M, Langaas S. Are agricultural soils dumps for microplastics of urban origin? *Environ Sci Technol* 2016;**50**:10777–9.
62. Scheurer M, Bigalke M. Microplastics in Swiss floodplain soils. *Environ Sci Technol* 2018;**52**:3591–8.
63. Coll C, Notter D, Gottschalk F, Sun T, Som C, Nowack B. Probabilistic environmental risk assessment of five nanomaterials (nano-TiO_2, nano-Ag, nano-ZnO, CNT, and fullerenes). *Nanotoxicology* 2016;**10**:436–44.
64. Pico Y, Andreu V. Analytical tools able to detect ENP/NM/MNs in both artificial and natural environmental water media. In: Julian Blasco IC, editor. *Ecotoxicology of nanoparticles in aquatic systems*. Boca Ratón: CRC Press; 2019. p. 230–59.
65. von der Kammer F, Ferguson PL, Holden PA, Masion A, Rogers KR, Klaine SJ, et al. Analysis of engineered nanomaterials in complex matrices (environment and biota): general considerations and conceptual case studies. *Environ Toxicol Chem* 2012;**31**:32–49.
66. Nowack B, Baalousha M, Bornhöft N, Chaudhry Q, Cornelis G, Cotterill J, et al. Progress towards the validation of modeled environmental concentrations of engineered nanomaterials by analytical measurements. *Environ Sci Nano* 2015;**2**:421–8.
67. Gogos A, Kaegi R, Zenobi R, Bucheli TD. Capabilities of asymmetric flow field-flow fractionation coupled to multi-angle light scattering to detect carbon nanotubes in soot and soil. *Environ Sci Nano* 2014;**1**:584–94.
68. Bauerlein PS, Emke E, Tromp P, Hofman J, Carboni A, Schooneman F, et al. Is there evidence for man-made nanoparticles in the Dutch environment? *Sci Total Environ* 2017;**576**:273–83.
69. Gottschall N, Edwards M, Topp E, Bolton P, Payne M, Curnoe WE, et al. Nitrogen, phosphorus, and bacteria tile and groundwater quality following direct injection of dewatered municipal biosolids into soil. *J Environ Qual* 2009;**38**:1066–75.
70. Gottschall N, Topp E, Edwards M, Payne M, Kleywegt S, Russell P, et al. Hormones, sterols, and fecal indicator bacteria in groundwater, soil, and subsurface drainage following a high single application of municipal biosolids to a field. *Chemosphere* 2013; **91**:275–86.
71. Liu HH, Cohen Y. Multimedia environmental distribution of engineered nanomaterials. *Environ Sci Technol* 2014;**48**:3281–92.
72. Christou A, Agüera A, Bayona JM, Cytryn E, Fotopoulos V, Lambropoulou D, et al. The potential implications of reclaimed wastewater reuse for irrigation on the agricultural environment: the knowns and unknowns of the fate of antibiotics and antibiotic resistant bacteria and resistance genes—a review. *Water Res* 2017;**123**:448–67.
73. Pino-Otín MR, Muñiz S, Val J, Navarro E. Effects of 18 pharmaceuticals on the physiological diversity of edaphic microorganisms. *Sci Total Environ* 2017;**595**:441–50.
74. EMEA. *Guideline on the environmental risk assessment of medicinal products for human use.* EMEA/CHMP/SWP/4447/00 corr 21European Medicine Agency, Committee for Medicinal Products for Human Use (CHMP); 2006.
75. Thomaidi VS, Stasinakis AS, Borova VL, Thomaidis NS. Assessing the risk associated with the presence of emerging organic contaminants in sludge-amended soil: a country-level analysis. *Sci Total Environ* 2016;**548–549**:280–8.
76. Martín J, Camacho-Muñoz D, Santos JL, Aparicio I, Alonso E. Occurrence of pharmaceutical compounds in wastewater and sludge from wastewater treatment plants: removal and ecotoxicological impact of wastewater discharges and sludge disposal. *J Hazard Mater* 2012;**239–240**:40–7.

77. Martín J, Camacho-Muñoz MD, Santos JL, Aparicio I, Alonso E. Distribution and temporal evolution of pharmaceutically active compounds alongside sewage sludge treatment. Risk assessment of sludge application onto soils. *J Environ Manage* 2012;**102**:18–25.
78. Nizzetto L, Bussi G, Futter MN, Butterfield D, Whitehead PG. A theoretical assessment of microplastic transport in river catchments and their retention by soils and river sediments. *Environ Sci: Processes Impacts* 2016;**18**:1050–9.
79. Cao D, Wang X, Luo X, Liu G, Zheng H. Effects of polystyrene microplastics on the fitness of earthworms in an agricultural soil. In: *IOP conference series: earth and environmental science*. IOP Publishing; 2017. p. 012148.
80. Huerta Lwanga E, Gertsen H, Gooren H, Peters P, Salánki T, van der Ploeg M, et al. Microplastics in the terrestrial ecosystem: implications for Lumbricus terrestris (Oligochaeta, Lumbricidae). *Environ Sci Technol* 2016;**50**:2685–91.
81. Huerta Lwanga E, Gertsen H, Gooren H, Peters P, Salánki T, van der Ploeg M, et al. Incorporation of microplastics from litter into burrows of Lumbricus terrestris. *Environ Pollut* 2017;**220**:523–31.
82. Koelmans AA, Bakir A, Burton GA, Janssen CR. Microplastic as a vector for chemicals in the aquatic environment: critical review and model-supported reinterpretation of empirical studies. *Environ Sci Technol* 2016;**50**:3315–26.
83. Huerta Lwanga E, Thapa B, Yang X, Gertsen H, Salánki T, Geissen V, et al. Decay of low-density polyethylene by bacteria extracted from earthworm's guts: a potential for soil restoration. *Sci Total Environ* 2018;**624**:753–7.
84. Rodriguez-Seijo A, Lourenço J, Rocha-Santos T, Da Costa J, Duarte A, Vala H, et al. Histopathological and molecular effects of microplastics in Eisenia andrei Bouché. *Environ Pollut* 2017;**220**:495–503.
85. Ng E-L, Lwanga EH, Eldridge SM, Johnston P, Hu H-W, Geissen V, et al. An overview of microplastic and nanoplastic pollution in agroecosystems. *Sci Total Environ* 2018;**627**:1377–88.
86. Wang J, Lv S, Zhang M, Chen G, Zhu T, Zhang S, et al. Effects of plastic film residues on occurrence of phthalates and microbial activity in soils. *Chemosphere* 2016;**151**:171–7.
87. Mortimer M, Holden PA. Fate of engineered nanomaterials in natural environments and impacts on ecosystems. In: Marmiroli N, White JC, Song J, editors. *Exposure to engineered nanomaterials in the environment*. Amsterdam: Elsevier; 2019. p. 61–103.
88. Hayat S, Pichtel J, Faizan M, Fariduddin Q. *Sustainable agriculture reviews 41 nanotechnology for plant growth and development*. Cham, Switzerland: Springer; 2020.
89. Tong Z, Bischoff M, Nies L, Applegate B, Turco RF. Impact of fullerene (C60) on a soil microbial community. *Environ Sci Technol* 2007;**41**:2985–91.
90. Johansen A, Pedersen AL, Jensen KA, Karlson U, Hansen BM, Scott-Fordsmand JJ, et al. Effects of C60 fullerene nanoparticles on soil bacteria and protozoans. *Environ Toxicol Chem* 2008;**27**:1895–903.
91. Mueller NC, Nowack B. Exposure modeling of engineered nanoparticles in the environment. *Environ Sci Technol* 2008;**42**:4447–53.
92. Gottschalk F, Nowack B. A probabilistic method for species sensitivity distributions taking into account the inherent uncertainty and variability of effects to estimate environmental risk. *Integr Environ Assess Manag* 2013;**9**:79–86.
93. Ghirardini A, Verlicchi P. A review of selected microcontaminants and microorganisms in land runoff and tile drainage in treated sludge-amended soils. *Sci Total Environ* 2019;**655**:939–57.
94. Gottschall N, Topp E, Metcalfe C, Edwards M, Payne M, Kleywegt S, et al. Pharmaceutical and personal care products in groundwater, subsurface drainage, soil, and wheat grain, following a high single application of municipal biosolids to a field. *Chemosphere* 2012;**87**:194–203.

95. Engdahl NB. Simulating the mobility of micro-plastics and other fiber-like objects in saturated porous media using constrained random walks. *Adv Water Resour* 2018;**121**:277–84.
96. Rillig MC, Ingraffia R, de Souza Machado AA. Microplastic incorporation into soil in agroecosystems. *Front Plant Sci* 2017;**8**:1805.
97. Galloway TS, Cole M, Lewis C. Interactions of microplastic debris throughout the marine ecosystem. *Nat Ecol Evol* 2017;**1**:0116.
98. Bläsing M, Amelung W. Plastics in soil: analytical methods and possible sources. *Sci Total Environ* 2018;**612**:422–35.
99. Yuan G, Theng BKG, Feng L, Bi D. Fate of engineered nanomaterials in urban and work environments. In: Marmiroli N, White JC, Song J, editors. *Exposure to engineered nanomaterials in the environment*. Amsterdam: Elsevier; 2019. p. 143–63.
100. Wang Y, Li Y, Kim H, Walker SL, Abriola LM, Pennell KD. Transport and retention of fullerene nanoparticles in natural soils. *J Environ Qual* 2010;**39**:1925–33.
101. Li Y, Wang Y, Pennell KD, Abriola LM. Investigation of the transport and deposition of fullerene (C60) nanoparticles in Quartz sands under varying flow conditions. *Environ Sci Technol* 2008;**42**:7174–80.
102. Zhang L, Hou L, Wang L, Kan AT, Chen W, Tomson MB. Transport of fullerene nanoparticles (nC60) in saturated sand and sandy soil: controlling factors and modeling. *Environ Sci Technol* 2012;**46**:7230–8.
103. Navarro DA, Kookana RS, Kirby JK, Martin SM, Shareef A, Du J, et al. Behaviour of fullerenes (C60) in the terrestrial environment: potential release from biosolids-amended soils. *J Hazard Mater* 2013;**262**:496–503.
104. Farré M, Sanchís J, Barceló D. Adsorption and desorption properties of carbon nanomaterials, the potential for water treatments and associated risks. In: Lofrano G, Libralato G, Brown J, editors. *Nanotechnologies for environmental remediation*. **2017**: Cham: Springer; 2017. p. 137–82.
105. Kinney CA, Heuvel BV. Translocation of pharmaceuticals and personal care products after land application of biosolids. *Curr Opin Environ Sci Health* 2020;**14**:23–30.
106. Christou A, Papadavid G, Dalias P, Fotopoulos V, Michael C, Bayona JM, et al. Ranking of crop plants according to their potential to uptake and accumulate contaminants of emerging concern. *Environ Res* 2019;**170**:422–32.
107. Pan M, Chu LM. Fate of antibiotics in soil and their uptake by edible crops. *Sci Total Environ* 2017;**599–600**:500–12.
108. Bartrons M, Peñuelas J. Pharmaceuticals and personal-care products in plants. *Trends Plant Sci* 2017;**22**:194–203.
109. Miller EL, Nason SL, Karthikeyan KG, Pedersen JA. Root uptake of pharmaceuticals and personal care product ingredients. *Environ Sci Tech* 2016;**50**:525–41.
110. Wu X, Dodgen LK, Conkle JL, Gan J. Plant uptake of pharmaceutical and personal care products from recycled water and biosolids: a review. *Sci Total Environ* 2015;**536**:655–66.
111. Goldstein M, Shenker M, Chefetz B. Insights into the uptake processes of wastewater-borne pharmaceuticals by vegetables. *Environ Sci Tech* 2014;**48**:5593–600.
112. Dodgen LK, Ueda A, Wu X, Parker DR, Gan J. Effect of transpiration on plant accumulation and translocation of PPCP/EDCs. *Environ Pollut* 2015;**198**:144–53.
113. Zhang Y, Sallach JB, Hodges L, Snow DD, Bartelt-Hunt SL, Eskridge KM, et al. Effects of soil texture and drought stress on the uptake of antibiotics and the internalization of Salmonella in lettuce following wastewater irrigation. *Environ Pollut* 2016;**208**:523–31.
114. Li Y, Sallach JB, Stephen WZ, Boyd SA, Li H. Insight into the distribution of pharmaceuticals in soil-water-plant systems. *Water Res* 2019;**152**:38–46.
115. Carter LJ, Williams M, Martın S, Kamaludeen SPB, Kookana RS. Sorption, plant uptake and metabolism of benzodiazepines. *Sci Total Environ* 2018;**628–629**:18–25.

116. Ahmed AM, Shimamoto T. Molecular analysis of multidrug resistance in Shiga toxin-producing Escherichia coli O157:H7 isolated from meat and dairy products. *Int J Food Microbiol* 2015;**193**:68–73.
117. Williams M, Martin S, Kookana RS. Sorption and plant uptake of pharmaceuticals from an artificially contaminated soil amended with biochars. *Plant and Soil* 2015;**395**:75–86.
118. Carter LJ, Harris E, Williams M, Ryan JJ, Kookana RS, Boxall ABA. Fate and uptake of pharmaceuticals in soil-plant systems. *J Agric Food Chem* 2014;**62**:816–25.
119. Herklotz PA, Gurung P, Vanden HB, Kinney CA. Uptake of human pharmaceuticals by plants grown under hydroponic conditions. *Chemosphere* 2010;**78**:1416–21.
120. Kodešová R, Klement A, Golovko O, Fér M, Kočárek M, Nikodem A, et al. Soil influences on uptake and transfer of pharmaceuticals from sewage sludge amended soils to spinach. *J Environ Manage* 2019;**250**:109407.
121. Wu C, Spongberg AL, Witter JD, Maruthi Sridhar BB. Transfer of wastewater associated pharmaceuticals and personal care products to crop plants from biosolids treated soil. *Ecotoxicol Environ Saf* 2012;**85**:104–9.
122. Holling CS, Bailey JL, Vanden Heuvel B, Kinney CA. Uptake of human pharmaceuticals and personal care products by cabbage (Brassica campestris) from fortified and biosolids-amended soils. *J Environ Monit* 2012;**14**:3029–36.
123. Wu X, Spongberg AL, Witter JD, Czajkowski KP. Uptake of pharmaceutical and personal care products by soybean plants from soils applied with biosolids and irrigated with contaminated water. *Environ Sci Tech* 2010;**44**:6157–61.
124. Mordechay EB, Tarchitzky J, Chen Y, Shenker M, Chefetz B. Composted biosolids and treated wastewater as sources of pharmaceuticals and personal care products for plant uptake: a case study with carbamazepine. *Environ Pollut* 2018;**232**:164–72.
125. Sabourin L, Duenk P, Bonte-Gelok S, Payne M, Lapen DR, Topp E. Uptake of pharmaceuticals, hormones and parabens into vegetables grown in soil fertilized with municipal biosolids. *Sci Total Environ* 2012;**431**:233–6.
126. Ahmed MBM, Rajapaksha AU, Lim JE, Vu NT, Kim IS, Kang HM, et al. Distribution and accumulative pattern of tetracyclines and sulfonamides in edible vegetables of cucumber, tomato and lettuce. *J Agric Food Chem* 2015;**63**:398–405.
127. Klampfl CW. Metabolization of pharmaceuticals by plants after uptake from water and soil: a review. *Trends Anal Chem* 2019;**111**:13–26.
128. Bártíková H, Skálová L, Stuchlíková L, Vokřál I, Vaněk T, Podlipná R. Xenobiotic-metabolizing enzymes in plants and their role in uptake and biotransformation of veterinary drugs in the environment. *Drug Metab Rev* 2015;**47**:374–87.
129. Macherius A, Eggen T, Lorenz W, Moeder M, Ondruschka J, Reemtsma T. Metabolization of the bacteriostatic agent triclosan in edible plants and its consequences for plant uptake assessment. *Environ Sci Tech* 2012;**46**:10797–804.
130. Riemenschneider C, Seiwert B, Moeder M, Schwarz D, Reemtsma T. Extensive transformation of the pharmaceutical carbamazepine following uptake into intact tomato plants. *Environ Sci Tech* 2017;**51**:6100–9.
131. Bartha B, Huber C, Schröder P. Uptake and metabolism of diclofenac in Typha latifolia—how plants cope with human pharmaceutical pollution. *Plant Sci* 2014;**227**:12–20.
132. Picó Y, Alvarez-Ruiz R, Wijaya L, Alfarhan A, Alyemeni M, Barceló D. Analysis of ibuprofen and its main metabolites in roots, shoots, and seeds of cowpea (*Vigna unguiculata* L. Walp) using liquid chromatography-quadrupole time-of-flight mass spectrometry: uptake, metabolism, and translocation. *Anal Bioanal Chem* 2018;**410**:1163–76.
133. Li Y, Chuang YH, Sallach JB, Zhang W, Boyd SA, Li H. Potential metabolism of pharmaceuticals in radish: comparison of in vivo and in vitro exposure. *Environ Pollut* 2018;**242**:962–9.

134. Bandmann V, Müller JD, Köhler T, Homann U. Uptake of fluorescent nano beads into BY2-cells involves clathrin-dependent and clathrin-independent endocytosis. *FEBS Lett* 2012;**586**:3626–32.
135. Li L, Zhou Q, Yin N, Tu C, Luo Y. Uptake and accumulation of microplastics in an edible plant. *Chin Sci Bull* 2019;**64**:928–34.
136. Wan Y, Wu C, Xue Q, Hui X. Effects of plastic contamination on water evaporation and desiccation cracking in soil. *Sci Total Environ* 2019;**654**:576–82.
137. Rillig MC, Lehmann A, de Souza Machado AA, Yang G. Microplastic effects on plants. *New Phytol* 2019;**223**:1066–70.
138. Ma C, White JC, Zhao J, Zhao Q, Xing B. Uptake of engineered nanoparticles by food crops: characterization, mechanisms, and implications. *Annu Rev Food Sci Technol* 2018;**9**:129–53.
139. Kole C, Kole P, Randunu KM, Choudhary P, Podila R, Ke PC, et al. Nanobiotechnology can boost crop production and quality: first evidence from increased plant biomass, fruit yield and phytomedicine content in bitter melon (*Momordica charantia*). *BMC Biotechnol* 2013;**13**:37.
140. Avanasi R, Jackson WA, Sherwin B, Mudge JF, Anderson TA. C60 fullerene soil sorption, biodegradation, and plant uptake. *Environ Sci Technol* 2014;**48**:1792–6.
141. Zhao Q, Ma C, White JC, Dhankher OP, Zhang X, Zhang S, et al. Quantitative evaluation of multi-wall carbon nanotube uptake by terrestrial plants. *Carbon* 2017;**114**:661–70.
142. Gogos A, Moll J, Klingenfuss F, van der Heijden M, Irin F, Green MJ, et al. Vertical transport and plant uptake of nanoparticles in a soil mesocosm experiment. *J Nanobiotechnol* 2016;**14**:40.
143. Carvalho PN, Basto MCP, Almeida CMR, Brix H. A review of plant-pharmaceutical interactions: from uptake and effects in crop plants to phytoremediation in constructed wetlands. *Environ Sci Pollut Res* 2014;**21**:11729–63.
144. Jjemba PK. The potential impact of veterinary and human therapeutic agents in manure and biosolids on plants grown on arable land: a review. *Agric Ecosyst Environ* 2002;**93**:267–78.
145. Hillis DG, Fletcher J, Solomon KR, Sibley PK. Effects of ten antibiotics on seed germination and root elongation in three plant species. *Arch Environ Contam Toxicol* 2011;**60**:220–32.
146. Liu F, Ying GG, Tao R, Jian-Liang Z, Yang JF, Zhao LF. Effects of six selected antibiotics on plant growth and soil microbial and enzymatic activities. *Environ Pollut* 2009;**157**:1636–42.
147. Kong WD, Zhu YG, Liang YC, Zhang J, Smith FA, Yang A. Uptake of oxytetracycline and its phytotoxicity to alfalfa (Medicago sativa L.). *Environ Pollut* 2007;**147**:187–93.
148. Migliore L, Cozzolino S, Fiori M. Phytotoxicity to and uptake of enrofloxacin in crop plants. *Chemosphere* 2003;**52**:1233–44.
149. Carter LJ, Williams M, Böttcher C, Kookana RS. Uptake of pharmaceuticals influences plant development and affects nutrient and hormone homeostases. *Environ Sci Tech* 2015;**49**:12509–18.
150. Feito R, Valcarcel Y, Catala M. Preliminary data suggest that venlafaxine environmental concentrations could be toxic to plants. *Chemosphere* 2013;**90**:2065–9.
151. Schmidt W, Redshaw CH. Evaluation of biological endpoints in crop plants after exposure to non-steroidal anti-inflammatory drugs (NSAIDs): implications for phytotoxicological assessment of novel contaminants. *Ecotoxicol Environ Saf* 2015;**112**: 212–22.
152. O'Connor S, Aga DS. Analysis of tetracycline antibiotics in soil: advances in extraction, clean-up, and quantification. *TrAC Trends Anal Chem* 2007;**26**:456–65.

153. Michelini L, La Rocca N, Rascio N, Ghisi R. Structural and functional alterations induced by two sulfonamide antibiotics on barley plants. *Plant Physiol Biochem* 2013;**67**:55–62.
154. Bártíková H, Podlipná R, Skálová L. Veterinary drugs in the environment and their toxicity to plants. *Chemosphere* 2016;**144**:2290–301.
155. Christou A, Antoniou C, Christodoulou C, Hapeshi E, Stavrou I, Michael C, et al. Stress-related phenomena and detoxification mechanisms induced by common pharmaceuticals in alfalfa (*Medicago sativa* L.) plants. *Sci Total Environ* 2016;**557–558**:652–64.
156. Bosker T, Bouwman LJ, Brun NR, Behrens P, Vijver MG. Microplastics accumulate on pores in seed capsule and delay germination and root growth of the terrestrial vascular plant Lepidium sativum. *Chemosphere* 2019;**226**:774–81.
157. Qi Y, Yang X, Pelaez AM, Lwanga EH, Beriot N, Gertsen H, et al. Macro-and microplastics in soil-plant system: effects of plastic mulch film residues on wheat (Triticum aestivum) growth. *Sci Total Environ* 2018;**645**:1048–56.
158. De la Rosa G, Garcia-Castaneda C, Vazquez-Nunez E, Alonso-Castro AJ, Basurto-Islas G, Mendoza A, et al. Physiological and biochemical response of plants to engineered NMs: implications on future design. *Plant Physiol Biochem* 2017;**110**:226–35.
159. Vithanage M, Seneviratne M, Ahmad M, Sarkar B, Ok YS. Contrasting effects of engineered carbon nanotubes on plants: a review. *Environ Geochem Health* 2017;**39**:1421–39.
160. Kumar V, Lakkaboyana SK, Sharma N, Abdelaal AS, Maitra SS, Pant D. Engineered nanomaterials uptake, bioaccumulation and toxicity mechanisms in plants. In: Verma SK, Das AK, editors. *Engineered nanomaterials and phytonanotechnology: Challenges for plant sustainability*. Amsterdam: Elsevier; 2019. p. 111–31.
161. Yuan Z, Zhang Z, Wang X, Li L, Cai K, Han H. Novel impacts of functionalized multi-walled carbon nanotubes in plants: promotion of nodulation and nitrogenase activity in the rhizobium-legume system. *Nanoscale* 2017;**9**:9921–37.
162. Hatami M. Toxicity assessment of multi-walled carbon nanotubes on Cucurbita pepo L. under well-watered and water-stressed conditions. *Ecotoxicol Environ Saf* 2017;**142**:274–83.
163. Roccaro P, Vagliasindi FGA. Risk assessment of the use of biosolids containing emerging organic contaminants in agriculture. *Chem Eng Trans* 2014;**37**:817–22.
164. Prosser RS, Sibley PK. Human health risk assessment of pharmaceuticals and personal care products in plant tissue due to biosolids and manure amendments, and wastewater irrigation. *Environ Int* 2015;**75**:223–33.
165. WHO. *Global action plan on antimicrobial resistance*. Geneva, Switzerland: WHO; 2015.
166. EC. *European comission, a European one health action plan against antimicrobial resistance (AMR)*. https://ec.europa.eu/health/amr/antimicrobial-resistance_en; 2017.
167. Zhang H, Li X, Yang Q, Sun L, Yang X, Zhou M, et al. Plant growth, antibiotic uptake, and prevalence of antibiotic resistance in an endophytic system of Pakchoi under antibiotic exposure. *Int J Environ Res Public Health* 2017;**14**:1336.
168. Bondarczuk K, Markowicz A, Piotrowska-Seget Z. The urgent need for risk assessment on the antibiotic resistance spread via sewage sludge land application. *Environ Int* 2016;**87**:49–55.
169. Chen Q, An X, Li H, Su J, Ma Y, Zhu YG. Long-term field application of sewage sludge increases the abundance of antibiotic resistance genes in soil. *Environ Int* 2016;**92-93**:1–10.
170. Riber L, Poulsen PH, Al-Soud WA, Skov Hansen LB, Bergmark L, Brejnrod A, et al. Exploring the immediate and long-term impact on bacterial communities in soil amended with animal and urban organic waste fertilizers using pyrosequencing and screening for horizontal transfer of antibiotic resistance. *FEMS Microbiol Ecol* 2014;**90**:206–24.

171. Rahube TO, Marti R, Scott A, Tien YC, Murray R, Sabourin L, et al. Impact of fertilizing with raw or anaerobically digested sewage sludge on the abundance of antibiotic-resistant coliforms, antibiotic resistance genes, and pathogenic bacteria in soil and on vegetables at harvest. *Appl Environ Microbiol* 2014;**80**:6898–907.
172. Burch TR, Sadowsky MJ, LaPara TM. Fate of antibiotic resistance genes and class 1 integrons in soil microcosms following the application of treated residual municipal wastewater solids. *Environ Sci Technol* 2014;**48**:5620–7.
173. Joy SR, Bartelt-Hunt SL, Snow DD, Gilley JE, Woodbury BL, Parker DB, et al. Fate and transport of antimicrobials and antimicrobial resistance genes in soil and runoff following land application of swine manure slurry. *Environ Sci Technol* 2013;**47**:12081–8.
174. Kelessidis A, Stasinakis AS. Comparative study of the methods used for treatment and final disposal of sewage sludge in European countries. *Waste Manag* 2012;**32**:1186–95.
175. Seleiman MF. Recycling sludge on cropland as fertilizer—advantages and risks. *Resour Conserv Recycl* 2020;**155**:104647.
176. Gonzalez-Gil L, Papa M, Feretti D, Ceretti E, Mazzoleni G, Steimberg N, et al. Is anaerobic digestion effective for the removal of organic micropollutants and biological activities from sewage sludge? *Water Res* 2016;**102**:211–20.
177. Gang X, Mingji J, Hong C, Min S, Yunfan L, Xiang L, et al. Critical review of ARGs reduction behavior in various sludge and sewage treatment processes in wastewater treatment plants. *Crit Rev Environ Sci Technol* 2019;**49**:1623–74.
178. Zhou H, Zhang Z, Wang M, Hu T, Wang Z. Enhancement with physicochemical and biological treatments in the removal of pharmaceutically active compounds during sewage sludge anaerobic digestion processes. *Chem Eng J* 2017;**316**:361–9.
179. Zhen G, Lu X, Kato H, Zhao Y, Li Y-Y. Overview of pretreatment strategies for enhancing sewage sludge disintegration and subsequent anaerobic digestion: current advances, full-scale application and future perspectives. *Renew Sustain Energy Rev* 2017;**69**:559–77.
180. Pei J, Yao H, Wang H, Ren J, Yu X. Comparison of ozone and thermal hydrolysis combined with anaerobic digestion for municipal and pharmaceutical waste sludge with tetracycline resistance genes. *Water Res* 2016;**99**:122–8.
181. Tong J, Lu X, Zhang J, Sui Q, Wang R, Chen M, et al. Occurrence of antibiotic resistance genes and mobile genetic elements in enterococci and genomic DNA during anaerobic digestion of pharmaceutical waste sludge with different pretreatments. *Bioresour Technol* 2017;**235**:316–24.
182. Wang M, Li R, Zhao Q. Distribution and removal of antibiotic resistance genes during anaerobic sludge digestion with alkaline, thermal hydrolysis and ultrasonic pretreatments. *Front Environ Sci Eng* 2019;**13**:43.
183. Zhang Z, Chen Y. Effects of microplastics on wastewater and sewage sludge treatment and their removal: a review. *Chem Eng J* 2020;**382**:122955.
184. Min Jang H, Choi S, Shin J, Kan E, Mo KY. Additional reduction of antibiotic resistance genes and human bacterial pathogens via thermophilic aerobic digestion of anaerobically digested sludge. *Bioresour Technol* 2019;**273**:259–68.
185. Tomei MC, Mosca Angelucci D, Mascolo G, Kunkel U. Post-aerobic treatment to enhance the removal of conventional and emerging micropollutants in the digestion of waste sludge. *Waste Manag* 2019;**96**:36–46.
186. Cucina M, Zadra C, Marcotullio MC, Di Maria F, Sordi S, Curini M, et al. Recovery of energy and plant nutrients from a pharmaceutical organic waste derived from a fermentative biomass: integration of anaerobic digestion and composting. *J Environ Chem Eng* 2017;**5**:3051–7.
187. Gurmessa B, Pedretti EF, Cocco S, Cardelli V, Corti G. Manure anaerobic digestion effects and the role of pre- and post-treatments on veterinary antibiotics and antibiotic resistance genes removal efficiency. *Sci Total Environ* 2020;**721**:137532.

188. Yu Z, Tang J, Liao H, Liu X, Zhou P, Chen Z, et al. The distinctive microbial community improves composting efficiency in a full-scale hyperthermophilic composting plant. *Bioresour Technol* 2018;**265**:146–54.
189. Zhang J, Sui Q, Lu T, Zhong H, Shen P, Wei Y. Sludge bio-drying followed by land application could control the spread of antibiotic resistance genes. *Environ Int* 2019;**130**:104906.
190. Chen Z, Zhao W, Xing R, Xie S, Yang X, Cui P, et al. Enhanced in situ biodegradation of microplastics in sewage sludge using hyperthermophilic composting technology. *J Hazard Mater* 2020;**384**:121271.
191. Pilnáček V, Innemanová P, Šereš M, Michalíková K, Stránská Š, Wimmerová L, et al. Micropollutant biodegradation and the hygienization potential of biodrying as a pretreatment method prior to the application of sewage sludge in agriculture. *Ecol Eng* 2019;**127**:212–9.
192. Zhang J, Sui Q, Tong J, Buhe C, Wang R, Chen M, et al. Sludge bio-drying: effective to reduce both antibiotic resistance genes and mobile genetic elements. *Water Res* 2016;**106**:62–70.
193. Wang L, Chang Y, Liu Q. Fate and distribution of nutrients and heavy metals during hydrothermal carbonization of sewage sludge with implication to land application. *J Clean Prod* 2019;**225**:972–83.
194. vom Eyser C, Palmu K, Schmidt TC, Tuerk J. Pharmaceutical load in sewage sludge and biochar produced by hydrothermal carbonization. *Sci Total Environ* 2015;**537**:180–6.
195. vom Eyser C, Schmidt TC, Tuerk J. Fate and behaviour of diclofenac during hydrothermal carbonization. *Chemosphere* 2016;**153**:280–6.
196. Ducey TF, Collins JC, Ro KS, Woodbury BL, Griffin DD. Hydrothermal carbonization of livestock mortality for the reduction of pathogens and microbially-derived DNA. *Front Environ Sci Eng* 2017;**11**:9.
197. Tasca AL, Puccini M, Gori R, Corsi I, Galletti AMR, Vitolo S. Hydrothermal carbonization of sewage sludge: a critical analysis of process severity, hydrochar properties and environmental implications. *Waste Manag* 2019;**93**:1–13.

CHAPTER SIX

Modeling tools for risk management in reclaimed wastewater reuse systems: Focus on contaminants of emerging concern (CECs)

Riccardo Delli Compagni[a], Marco Gabrielli[a], Fabio Polesel[b,c], Andrea Turolla[a], Stefan Trapp[c], Luca Vezzaro[c], and Manuela Antonelli[a,]*

[a]Department of Civil and Environmental Engineering (DICA), Politecnico di Milano, Milano, Italy
[b]DHI A/S, Hørsholm, Denmark
[c]DTU Environment, Technical University of Denmark, Kongens Lyngby, Denmark
*Corresponding author: e-mail address: manuela.antonelli@polimi.it

Contents

1. Introduction 182
2. Reclaimed water reuse system: Boundaries and characteristics 185
   2.1 Constituents in reclaimed wastewater 185
3. Modeling reclaimed wastewater reuse systems for agricultural irrigation 190
   3.1 Why modeling a reclaimed wastewater reuse system 190
   3.2 Elements of the system and their conceptualization 192
   3.3 Fate processes 198
   3.4 Available modeling tools 203
4. Case studies 207
   4.1 CECs fate prediction through indirect RWW reuse and sludge application 207
   4.2 Predicting site-specific CECs consumption and health risk due to direct RWW reuse 209
5. Conclusions 213
References 213

Abstract

Reuse of reclaimed wastewater (RWW) for agricultural irrigation is becoming an essential practice in many arid and semi-arid regions of the world to save fresh water. Additionally, RWW contains essential plant nutrients (e.g., P, N, K), thus limiting the use of chemical fertilizers by farmers, who in the next years would need to substantially intensified their agriculture capacity to meet the increasing food demand. However, as consequences of human activities, other chemicals such as pharmaceuticals and

Advances in Chemical Pollution, Environmental Management and Protection, Volume 6
ISSN 2468-9289
https://doi.org/10.1016/bs.apmp.2020.07.010

biocides, currently referred to as "contaminants of emerging concern (CECs)," end up in RWW and subsequently in edible part of the crops, posing a risk for the environment and human health. This chapter aims at illustrating how modeling tools can help water managers in the identification of both effective monitoring campaigns and optimal RWW management strategies to minimize risk associated to conventional pollutants and especially to CECs. In fact, models can support the selection among the many hundred chemicals present in wastewater those of highest concern, but also address proper barrier identification once that a desired tolerable risk is set for specific CECs. The chapter illustrates how a RWW reuse system can be conceptualized into mathematical models, with a level of detail allowing to fully understand the processes taking place in each element of the system (wastewater treatment plant, distribution network, soil/plant) and to correctly interpret the simulation results. Two practical examples are also reported, showing the advantages of using modeling tools in risk management in complex RWW reuse systems.

Keywords: Indirect water reuse, Direct water reuse, Risk assessment, Scenario analysis, Modeling, Conceptualization, Wastewater, Contaminant of emerging concerns, Pharmaceuticals, PFAS

1. Introduction

A still growing human population requires increasing food supply. FAO estimates that by 2030 the world will require 60% more food, and that about 80% of that will be provided by intensifying agriculture through irrigation.[1] Irrigated agriculture depends on an adequate supply of water. As the easily available water sources are already widely exploited, agriculture needs water supply from new sources. Irrigation can rely on water with lower quality compared to drinking water supply, creating the opportunity for exploiting alternative water sources. Reuse of reclaimed wastewater (RWW) for agricultural irrigation limits the exploitation of high quality fresh water resources, not competing with other usages such as drinking water supply. This practice also enhances the provision of essential plant nutrients (e.g., P and N) still contained in RWW, which otherwise would be discharged into rivers and oceans, increasing eutrophication. In order to ensure health of plants, high growth yields and to avoid human health impacts, minimum quality standards were suggested in the past.[2,3] These existing recommendations were prepared three decades ago and focused on salinity, nutrient supply and conventional toxins, such as toxic elements and heavy metals. Meanwhile, new knowledge has raised awareness on new problematic constituents, which are not covered in these existing guidelines.

Many of these constituents, such as pharmaceuticals and biocides, are currently referred to as "contaminants of emerging concern" (CECs),[4]

which in the last years have garnered much attention from scientists due to their widespread occurrence in surface water and their potentially deleterious effects in aquatic life at low concentrations (<1 μg/L). Traditionally, CECs have not been recognized as a major concern since they were not found in concentrations above detection limits. However, the development of both new sophisticated analytical techniques and standards now allows detecting and monitoring them even at low ng/L levels.[5] These advances have highlighted CECs presence in sensitive environmental compartments, such as soil and edible parts of the crops, in some cases even posing a risk for the environment and human health.[6] On the other hand, the presence of nutrients, which are defined as "conventional contaminants" when discharged to natural water, becomes an advantage when considering RWW for crop irrigation. However, an excess of these substances can cause oversupply (reduced quality of the harvest product or reduced yields) of some crops,[3,7] thus requiring a proper management and experience.

In a RWW reuse system (i.e., wastewater treatment plants—WWTPs, surface water recipients, agro-ecosystems, see Fig. 1), treating wastewater to comply with drinking water standards may ensure a negligible risk from contaminants ("zero-risk" approach), be it conventional or emerging, but this is costly and exaggerated for non-potable reuse applications, where safety for workers and consumers is already guaranteed by existing, less stringent standards. As proposed in the guidelines for the safe use of wastewater, excreta and greywater by the World Health Organization,[8] state members should derive differentiated quality standards based on the "fit-for-purpose" principle. For example, standards for irrigation of crops intended for raw consumption should be stricter than those intended for products to be processed. Additionally, the formulation of quality standards should strongly

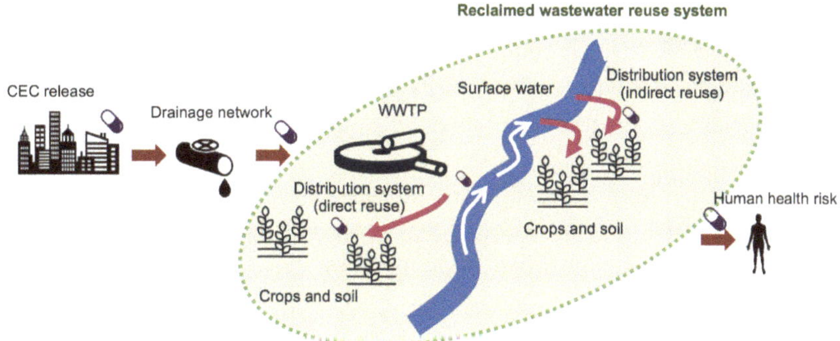

Fig. 1 Elements of the RWW reuse system in case of direct reuse and indirect reuse.

encourage the reuse of RWW, considering socio-cultural, economic and environmental conditions of the local communities, and the potential co-benefits created by an increase in the availability of water resources. The guidelines also state the need of using risk analysis to identify health-based criteria (e.g., treatment performances and water quality standards) and ensure human health protection. In particular, quantitative microbial (and chemical) risk assessment—QM(C)RA—is necessary to: (i) identify worst-case scenarios and sources of risk in order to prioritize control actions, and (ii) gain a perspective on the different nature and sources of the identified risks. Based on these principles, the 2018 EU regulation proposal for minimum water quality requirements[9] identified general quality standards for physical-chemical parameters and pathogen indicators for various RWW reuse scenarios, while leaving to risk analysis the role of deriving stricter or additional health-based criteria for case-specific hazards (e.g., CECs).

Assessing the existence of a risk requires understanding if chemical exposure concentrations exceed a tolerable threshold. Thresholds can be fixed by ecotoxicological studies or field observations, while exposure concentrations can be determined by either measurements or model predictions. Monitoring of thousands of CECs in complex RWW reuse systems can be both costly and time-consuming (e.g., using non-target screening[10]). As for nutrients, while their monitoring is feasible, their measurements tend to be discontinuous in time, hindering a comprehensive evaluation of the system's status. Quantitative modeling tools can help to overcome this lack of knowledge by providing valuable complementary information (e.g., filling temporal gaps between measurements), and accounting for the influence of site-specific conditions. Moreover, statistical indicators can be calculated from the predicted exposure concentrations, supporting a more robust risk quantification by, e.g., estimating probability of exceedance of certain thresholds. Traditional CECs modeling efforts have proven to be limited for complex RWW reuse systems, as they focused on: (i) single elements of the RWW reuse system[11–13] and, when evaluating integrated systems, the agro-ecosystem was neglected[14,15]; (ii) predicting nutrients/conventional contaminant dynamics (e.g., activated sludge models, river water quality models). Moreover, the conceptualization of the whole RWW reuse system into mathematical models requires experts with different backgrounds (e.g., agronomists, engineers, biologists, statisticians, etc.), which make its application still limited to few cases.

This chapter provides a practical guideline for researchers and practitioners aiming at developing mathematical models of RWW reuse systems and using them as decision-support tools for the identification of optimal water management strategies. The main goal is to facilitate model construction depending on the type of target contaminant to be controlled, with a special focus on CECs. In this chapter, models from the ASM-family[16] are highlighted as the most widespread tools at international level to predict the fate of nutrients. Besides, the most important processes for the majority of CECs are illustrated in form of differential equations, which can be easily integrated in the ASM models to provide quantitative results (e.g., concentration dynamics over a certain time interval) in each part of the RWW reuse system. This allows models to support two other important applications: (i) evaluation of environmental and human health chemical risk (QCRA), (ii) development of cost-effective monitoring programs to capture contaminant dynamics along RWW reuse systems.

2. Reclaimed water reuse system: Boundaries and characteristics

Based on the fit-for-purpose principle,[8] it is important to consider all the elements of a RWW system and their potential interactions (Fig. 1), since these are responsible for the environmental impacts and/or health risks. Two types of RWW reuse are possible: (i) direct reuse, which consists of introducing RWW directly into the agro-ecosystems, and (ii) indirect reuse, also known as de facto reuse or unplanned reuse, which refers to the introduction of RWW into an environmental buffer (e.g., surface water recipients) before being used in the agro-ecosystems.

2.1 Constituents in reclaimed wastewater

2.1.1 Macronutrients

RWW conveys important nutrients for plant growth. In most cases, the presence of nutrients (e.g., N, P and K) has been an advantage for farmers due to their importance for plant growth. Specifically, RWW can supply on average 225 kg of N and 45 kg of P per hectare per year, thus limiting the use of chemical fertilizers.[8] However, too high concentrations of nutrients can also be a disadvantage, leading to crop oversupply and, consequently, reduced yields and environmental impacts. The amount of optimal nutrient content in the irrigation water depends on the crop type, the amount of irrigation water, the time of the year, the soil type, additional nutrient sources and other, often site-specific, factors. For example, most crops are unaffected by nitrate in wastewater up to 30 mg N/L, while sensitive crops

Table 1 Main important nutrients for plant growth and their typical concentration in RWW, in comparison to common solutions used for plant growth.

Element	Typical range in raw wastewater (mg/L)	Typical range in tertiary treated RWW (mg/L)	Hoagland solution for hydroponics (mg/L)	ISO 8962 nutrient solution for willow tree tests [i] (mg/L)
Nitrogen (N)	20–70[a]	2–12[a]	210	40
Phosphorus (P)	4–12[a]	0.2–6[a]	31	7.2
Calcium (Ca)	42[b]–74[a]	17–70[a]	200	4.88
Magnesium (Mg)	11[b]–38[a]	16–40[a]	48.60	7.15
Potassium (K)	10–30[h]	6–26[a]	235	9.17
Sulfur (S)	14[b]–21[b]	13[b]–72[b]	64	1.95
Sodium (Na)	91[b]–418[c]	77–415[a]	1.2	65 (optional as buffer)
Iron (Fe)	0.008[a]–4[b]	0.05[b]	2.9	0.02
Boron (B)	0.35[a]	0.3[a]	0.5	0.03
Manganese (Mn)	0.06[a]–0.1[b]	0.002[a]	0.5	0.1
Zinc (Zn)	0.08[a]–2.4[b]	0.002[a]–0.05[a]	0.05	0.001
Copper (Cu)	0.06[a,d]–0.30[e]	0.006[g]–0.009[a]	0.02	<0.01
Nickel (Ni)	0.003[b]–0.040[f]	0.002[g]–0.004[a]	traces	none
Molybdenum (Mo)	traces[a]	traces[a]	traces	0.003

[a]17, [b]18, [c]19, [d]20, [e]21, [f]22, [g]23, [h]7, [i]24, 25.

(sugar beets, grapes) have a reduced quality of the harvest product even at much lower N content.[3] The nutrient demand of crops depends largely on the plant development, with plants in the exponential growth phase having the largest demands. The widely used Hoagland nutrient solution has a higher content of macronutrients (N, P, K) than most raw wastewaters (Table 1), but the values of the ISO standard solution 8692 are rather close to typical effluent concentrations of macro- and micro-nutrients, which means that WWTP effluents provide more or less all nutrients plants need for their growth. An excessive use of nutrients can also trigger leaching of nitrate to groundwater and washoff of P caused by surface runoff. These processes can potentially contaminate drinking water sources and contribute to eutrophication of surface water bodies. Consequently, control actions should be taken by farmers and water managers to adjust nutrient quantities before application, with different efforts depending on the type of RWW and crops. For example, the usage of fertilizers could be advantageously reduced or even eliminated in some cases by using RWW,[26] or optimal nutrient concentrations can be reached by blending RWW with other water sources. In conclusion, fit-to-purpose standards are necessary also for nutrients, taking into account the reuse system, the nutrient requirements of the irrigated crops, the availability of nutrients already present in soil and other site-specific conditions.

2.1.2 Salt

Salinity, the amount of salt dissolved in water, varies both in content and composition with origin and usage of water. Marcar et al.[27] list typical wastewater constituents contributing to salinity: inorganic nitrogen, phosphorous, sodium, calcium, magnesium, chloride and sulfate, with total dissolved solids (TDS) equal to 675 mg/L (hereof 135 mg Cl/L). The range of TDS is from 145 to 1224 mg/L, whereby 150 to 380 mg/L are added due to domestic water use. Salt non-tolerant species, like willows, show severe toxic effects when grown in hydroponic solution with >800–1000 mg/L sodium chloride, corresponding to 485–600 mg Cl/L.[28] Irrigation water with chloride levels above 350 mg Cl/L can be used only with restrictions, lower levels are required for sensitive crops.[17]

Plants reduce osmotic stress induced by elevated salt content of the external water with excretion of sodium chloride from root cells.[28–30] Hence, when water is taken up and salt is excluded, the salinity of the root zone inevitably increases with time, until it reaches phytotoxic levels, unless the salt is washed out by a surplus of water.[27,31] False irrigation schemes combined with salty water are, on a global scale, reason number one for degradation of once fertile soils. Presently, about 831 million hectares of the world's approximately 5 billion hectares, i.e., 16.6% of all agricultural land, are affected by salinization,[32] making salinization one of the major environmental problems. The removal of salt from water is rather costly, but wastewater can be mixed with other, less saline water sources. An alternative is the cultivation of salt tolerant cash crops.[33]

2.1.3 Micronutrients and heavy metals

As for the trace elements, some are essential for plant growth but may act toxic in higher quantities. When too high concentrations of some trace elements are present in RWW, their natural accumulation in edible parts of the crops can reach harmful concentrations, causing quality issues and posing a human health risk for consumers. For example, common elements in RWW, such as heavy metals (i.e., Cu and Zn) and boron, play all an important role in plant metabolism, but problems such as plant toxicity, leaf burn, premature leaf drops and reduced growth can occur when too high concentrations are present.[7] As for the human health risk, ingestion of contaminated food by heavy metals can lead to health problems including lung and skin diseases, kidney and neurological disorders, stomach problems, and also cancer.[34,35] The EU has set quality standards in food for a couple of toxic elements, among them lead and cadmium.[36] Table 1 summarizes the most important nutrients and trace elements for plant growth and their typical concentrations in RWW.

2.1.4 CECs

The term CECs refers to chemicals that may have been present in the environment for long time, but have only recently come into focus due to improving analytical methods and their negative effects on environment and human health.[37] A wide range of diffuse and point sources (e.g., highly populated cities, industries, agriculture, landfills, etc.) are responsible for introducing CECs into the environment through different pathways, including but not limited to treated wastewater as one of the main transport media. Most commonly detected CECs include natural and synthetic hormones and pharmaceuticals, such as analgesic and anti-inflammatory drugs, antibiotics, diuretics, psychiatric drugs, antidiabetics and antihistamines, with concentrations in wastewater influent generally varying from 1 ng/L to 1000 μg/L[38] (Table 2). Removal in conventional WWTPs varies significantly between substances and even for the same substance[38,43] and is mainly associated to biodegradation and sorption[44] and to a lesser extent volatilization and abiotic degradation. Negative removal efficiencies have been observed for some substances, such as diclofenac, carbamazepine and sulfonamide antibiotics. Possible explanations include deconjugation of conjugated metabolites, which are by-products produced by the human metabolization of the original molecule, and direct release from fecal particles.[38] However, inadequate sampling methods and analytical uncertainties can also explain negative values,[15,45–47] especially in large WWTPs.

Table 2 Typical concentrations of CECs at the inlet and outlet of European WWTPs.

CECs	Concentrations in WWTP influent (μg/L)	Concentrations in WWTP outlet (μg/L)
Antibiotics		
Clarithromycin	$0.40\cdot10^{-3}$–0.65[a]	$2.50\cdot10^{-2}$–0.36[a]
Sulfamethoxazole	<LOQ–11.56[a]	<LOQ–0.54[a]
Nonsteroidal anti-inflammatory drugs		
Ibuprofen	<LOQ–83.50[a]	<LOQ–24.60[a]
Paracetamol	<LOQ–482.69[a]	<LOQ–24.53[a]
Diclofenac	<LOQ–4.87[a]	<LOQ–5.16[a]
Diuretics		
Furosemide	0.17–1.08[b]	0.12–1.85[b]
Carbamazepine	<LOQ–3.11[a]	<LOQ–4.60[a]
Lipid regulators		
Gemfibrozil	<LOQ–17.06[a]	<LOQ–5.23[a]
Cardiovascular		
Atenolol	<LOQ–33.11[a]	<LOQ–7.60[a]
Antidiabetics		
Metformin	18.00–82.00[c]	1.20–26.00[c]
Antihistamines		
Ranitidine	$3.20\cdot10^{-2}$–0.23[b]	<LOQ–0.17[b]
Hormones		
Estrone	$2.40\cdot10^{-2}$–0.067[a]	<LOQ–$9.50\cdot10^{-2}$[a]
17α-ethynylstradiol	$0.40\cdot10^{-3}$–$7.00\cdot10^{-2}$[a]	$0.50\cdot10^{-3}$–0.11[a]
17β-estradiol	$2.50\cdot10^{-3}$–$5.20\cdot10^{-2}$[d]	<LOQ–$7\cdot10^{-3}$[d]
Perfluorinated compounds		
PFOS	<LOQ–0.45[e]	<LOQ–0.30[e]
PFOA	<LOQ–0.51[e]	<LOQ–0.47[e]

[a]39, [b]5, [c]40, [d]41, [e]42.

Advanced tertiary treatments, such as activated carbon and ozone, have shown comparably high removal efficiencies (i.e., >90%) for a broad range of pharmaceuticals.[43] However, the use of these technologies has been associated to a number of drawbacks, such as the high costs and the potential formation of hazardous by-products, and optimal advanced treatment trains in view of RWW reuse have been recently identified.[48] To date, few countries (Switzerland, Germany, Sweden) are systematically implementing advanced treatment in municipal WWTPs,[49] mainly due to the high costs and the absence of a specific regulation for CECs discharge.

Incomplete pharmaceutical removal in conventional WWTPs results in their occurrence in RWW, with potential subsequent translocation to edible plant parts after irrigation.[50] Consequently, CECs exposure through ingestion of irrigated food was observed for consumers, for whom non-negligible concentrations of some persistent CECs (i.e., carbamazepine) were recently found in urine.[51] While acute health effects seem unlikely, since concentration levels were four orders of magnitude lower than typical therapeutic concentrations, chronic effects (also due to multiple exposure to many substances) may not be excluded.

A category of CECs that has recently raised to prominence due to the ubiquitous presence in the environment and detrimental effects to living organisms are per- and polyfluoroalkyl substances (PFAS).[52] These substances are generally very recalcitrant to biodegradation, while sorption to sludge and soil is largely substance-dependent.[53] As a result of their comparably high persistence and mobility, PFASs are typically detected in RWW[42] and exhibit bioaccumulation potential in crops.[53] The presence of PFAS in agro-ecosystems has also been associated to their occurrence in sewage sludge[54] and to irrigation with contaminated groundwater.[55] The most commonly detected compounds in this category are perfluorooctanoic acid (PFOA) and perfluorooctanesulfonic acid (PFOS), for which considerable uptake in irrigated crops has been observed.[56,57]

3. Modeling reclaimed wastewater reuse systems for agricultural irrigation

3.1 Why modeling a reclaimed wastewater reuse system

3.1.1 Chemical risk assessment

As for quantitative tools in chemical risk assessment, models can be used in two manners: (i) as forward prediction tools, in which model inputs (e.g., consumption) and parameters (e.g., biokinetic coefficients) are known a priori and propagated through the model, providing predictions on CECs in target environmental compartments and to select among the many hundred chemicals present in wastewater those of highest concern (priority setting); (ii) as backward prediction tools for barrier identification once that a desired tolerable risk (for the environment and/or human health) is set for specific CECs.

3.1.2 Scenario analysis

Once that a RWW reuse system has been conceptualized into a mathematical model, scenario analysis can help to understand the effects of a certain action on the system performance. Examples of questions that forward prediction can address are "what will happen to the CECs concentrations in crops using a more efficient irrigation system?," "would the concentration decrease below safety thresholds, and would so the associated human health risk?." This will allow for the identification of optimal solutions in terms of cost-benefit and risk reduction, avoiding trial and error approaches on large full-scale systems. Indeed, optimal solutions can be found among all the multiple scenarios that can be run by a model.

3.1.3 Identification of optimal monitoring plans

Initially integrated in the Stockholm Framework and now in the European regulation proposal, the development of monitoring plans play an important role in the risk management of a water reuse system. Three main objectives have to be accomplished through monitoring: (i) proving that the system is capable of meeting the design requirements (validation monitoring), (ii) providing real-time information about the functionality of system components (operational monitoring), and (iii) verifying that health-based criteria are met at the points of use (verification monitoring). Operational monitoring typically relies on simple measurements (e.g., turbidity) in order to take quick decisions to prevent hazard outbreaks. Conversely, validation and especially verification monitoring typically require microbial and chemical laboratory analyses of samples collected at the point of use (e.g., reclaimed wastewater and crops). For this reason, the number of collected samples has to be optimized by identifying a priori location and timing of sampling in order to minimize costs and time of analyses while ensuring a proper knowledge of the contamination levels in the system to decision makers. Models can help to provide such a priori information, making prediction on the environmental compartments where specific CECs are more likely to be found. Moreover, models may also estimate the timing of peak concentrations, supporting the evaluation of worst-case situations.

3.2 Elements of the system and their conceptualization

The model structure should be defined according to the modeling goals, the components to be simulated and the data availability. Therefore, it is not necessary to reproduce each aspect of reality, but rather only the elements of the system and the processes that are relevant for the intended model applications and simulated components. Irrelevant processes are therefore discarded to ensure a parsimonious model structure.

In the following sections the conceptualization of each element of a reuse system and the data required for its simulation are described. Specifically, three elements of a reuse system have been identified: WWTP, distribution system and soil/plant compartment. Other elements could be considered/ present in a reuse system (e.g., sewer, other water sources), and their role with respect to the model scope should be further investigated.

3.2.1 Wastewater treatment plants

WWTPs are composed of an array of physical and biochemical unit processes to remove the pollutants present in wastewater. A series of hydrodynamic, physical, biological and physical-chemical processes occurs in each unit process, also creating interactions going beyond the system boundaries (e.g., due to aeration).

Depending on the scope of the simulations the description of specific processes might be required or be neglected (e.g., chemical precipitation, enhanced biological phosphorus removal, etc.). Rieger et al.[58] provide a guide on the selection of such processes and their details. Photolysis can be discarded from a WWTP model due to the high turbidity present in most tanks. On the contrary, volatilization and stripping are favored by the high mixing and the insufflation of air in use. Nonetheless, both these two processes are often negligible in case of CECs or metals simulation (with the exception of Hg), because these are frequently present as ions, which do not tend to volatilize.

As for biochemical processes in reaction tanks, mass transfer limitations can be neglected when dealing with suspended biomass, due to the intense mixing. However, in case of attached biomass, the mass transfer limitations between the bulk-water and the biofilm can play an important role. Indeed, it affects the substrate availability across the biofilm, possibly limiting nitrification, denitrification or biodegradation rates.[59,60] The high biomass concentration and the variable redox conditions in biological reactors promote the rate of the biochemical processes, such as nutrient removal and organic carbon reduction, and the sorption, biodegradation and deconjugation of CECs.[61,62] It is also important to stress that biodegradation, nitrification and denitrification do not exclusively occur within biological reactors, but also in other unit processes designed for other functions. For example, the use of a reactive settler model has shown to affect the effluent concentration of nutrients by about 10%,[63] and the return and waste sludge concentration of highly biodegradable CECs by about 100%, even though resulting in negligible changes in the effluent CECs concentrations.[64] The major removal process for certain CECs with limited biodegradability can be sorption and the subsequent solid-liquid separation.

An extensive data collection is required to collect information regarding the WWTP layout, influent and effluent characteristics, allowing to build, calibrate and validate the WWTP model (Table 3). In case site-specific experiments are not feasible, the influent concentrations of traditional pollutants can be estimated by using standard values from literature, as the ones present in Henze et al.,[59] or by model estimates as influent generators.[65] CECs influent pattern estimation, especially in case of biodegradable substances, might require also a conceptual description of the sewer system, as shown by Delli Compagni et al.[66] Furthermore, it is important to highlight that the influent composition might change significantly during the year, not only with respect to conventional pollution parameters, but also

Table 3 Data required for the setup, validation and calibration of a WWTP model.

Sub-model or task	Data required
Water quantity sub-model	– Process scheme – Number of lines – Volume and number of tanks – Flow rates (influent, effluent and between tanks)
Water quality sub-model	– Physical-chemical characteristics (e.g., temperature) – Pollutants concentrations – Operational set-points (e.g., tank oxygen concentration) – Kinetic model parameters incl. degradation rates – CECs physical-chemical parameters
Calibration, validation/verification	– Solids production and concentrations in tanks – Pollutants and nutrients concentrations in tanks or effluent – Nutrients concentrations in tanks or effluent – Sludge blanket height

to CECs type and concentrations, requiring a representative influent input when running long-term simulations. More details on the collection of data, model building, calibration and validation/verification can be found in the Good Modeling Practice guidelines described by Rieger et al.[58]

3.2.2 Distribution system

RWW has to be delivered from the WWTP outlet to the nearby fields: while existing canals or rivers are often used to deliver RWW, piping networks are also in use, depending if the reuse is direct or indirect. Similarly to WWTP modeling, distribution systems models have to describe hydrodynamic, biochemical, and physical processes affecting water quality during transport.

Piping distribution networks are described as graphs, where links represent the pipes and nodes represent the various demand points, storage tanks or junctions between pipes. While the water in pipes is subjected to advection and diffusion, nodes and storage tanks provide the possibility for mixing. Two compartments are usually addressed in pipes: the bulk-water and the pipe-inner-wall surface. The mass transfer between the two compartments is caused by sorption and desorption of dissolved chemicals, and attachment and detachment of particles.[67] Specific applications might also include the accumulation of loose sediments within pipes and an overhead space in tanks.

Rivers/canals can be described by using 2- or 3-dimensional model structures, but given the typical geometry ($3 \times 30 \times 30{,}000\,\text{m}$), one-dimensional models using advection-dispersion equations or a cascade of completely stirred tank reactors (CSTRs) are most widely applied, as complete mixing can be assumed following an initial mixing length.[68] The mixing of RWW with water of other sources might dilute some contaminants or, on the contrary, add other chemicals that will be delivered to the fields. Furthermore, depending on the specific scenario, it is possible that other sources, different from WWTPs, might impact the distributed water. Two compartments are usually addressed in river/canals: bulk-water and sediments, which are connected by diffusion of dissolved chemicals, and settling and re-suspension of particles. Sediments play a big role in the fate of CECs, as they tend to accumulate and store highly sorbable and weakly biodegradable CECs, which can then desorb in case of lower concentration in the bulk-water (e.g., due to a WWTP upgrade, or rainfall-induced increase in baseflow) or be transported downwards during periods of flooding.

The differences between distribution systems define the processes affecting RWW quality during transport. Pipes provide an environment that is less influenced by environmental variables than natural rivers/canals, and water quality is thereby only affected by: (i) nitrification/denitrification, and biodegradation, as bacteria are present even in the case of disinfection[69]; (ii) disinfectant degradation and disinfection by-product formation[70]; and (iii) sorption/desorption.[71] A distribution system comprising natural rivers/canals is additionally affected by volatilization, water re-aeration and photodegradation, which can also play a significant role in the chemicals fate.[72] CECs deconjugation can be neglected in distribution models, as this process can be considered completed during WWTP treatment.

Distribution system modeling requires, as WWTP modeling, an extensive data collection, including geometrical characteristics of the system, water abstraction, environmental variables, and pollutant concentrations (Table 4). One key aspect for simulating the concentration of chemicals delivered to the fields, especially in river/canals distribution systems, is the inclusion of daily and seasonal variations of environmental conditions (such as flowrate and quality of upstream water). In fact, these variations strongly affect the fate processes occurring during transport.

Table 4 Data required for the setup, validation and calibration of a RWW distribution model.

Sub-model or task	**Data required**
Water quantity sub-model	– Network layout and geometric characteristics (e.g., length, depth/diameter) – Location, periods and amount of abstraction – Upstream water flowrate (for indirect reuse)
Water quality sub-model	– WWTP effluent physical-chemical characteristics (e.g., temperature) – Pollutants concentrations in WWTP effluent – Kinetic model parameters incl. degradation rates – CECs physical-chemical parameters – Environmental variables (e.g., sunlight intensity, wind speed) – Upstream water pollutant concentration (for indirect reuse)
Calibration, validation/verification	– Water travel time – Pollutants concentrations at a downstream location

3.2.3 Soil and crops

As soil and crops influence heavily each other, their simulation cannot be carried out independently, but only within a single model including soil and crop sub-models. Soil is a heterogeneous medium whose composition differs along the soil column. For this reason, the soil sub-model cannot consider homogeneous conditions with depth, but it has to account for such variability. This is usually done by using different soil layers, characterized by different properties, possibly imitating the sequence of soil horizons present at the site. However, model simulations often only consider the plowed top layer of the soil as one well-mixed compartment.

Precipitation and irrigation provide water and solutes to the soil surface. Once infiltrated, water and dissolved chemicals move downwards or upwards due to percolation and capillary transport. These transport mechanisms can be described through advection-dispersion equations or with conceptual models, which represent the soil as a cascade of separate "tipping buckets."[68] While the latter description provides a solution with sufficient accuracy for field conditions, it does not describe complex flow processes (e.g., sub-surface drains). Hence, the plant uptake model is also connected to the soil solute transport model HYDRUS, which solves the Richards-equation.[73] The root water uptake, i.e., the transpiration, depends on the size and the photosynthesis or growth of the crops and it is often calculated by the plant sub-model.[74]

Soils contain diverse microbial communities, which can perform both nitrification and denitrification, depending on the oxic conditions. Furthermore, these bacteria are able to biodegrade dissolved organic chemicals, including many CECs. It is important to highlight that biodegradation rates vary extremely among sites and, for this reason, the use of a single biodegradation value from another site is not recommended, favoring instead the use of several different values in, e.g., a Monte-Carlo approach.[75]

Crop modeling does not refer to the growth and development of single plants, but consider a functional unit ground area with homogeneous plant and soil characteristics. As crops development is highly influenced by the site environmental conditions and water and nutrients availability, the role of the crop growth model is to link these conditions to the expected yields and assess the effect of unfavorable conditions.

When modeling pollutants accumulation in plants, the focus shifts from predicting the crops growth to estimate the expected concentration of a given pollutant in the different organs (roots, leaves, stem and fruits). The different organs can be modeled as different compartments connected by the phloem and xylem flows which carry the uptaken CECs. Via transpiration stream in the xylem, CECs are transported from roots to leaves. Differently, phloem tubes conduct CECs from the leaves to the other growing organs. Not all molecules present in soil can be uptaken: roots act as a barrier for molecules with molecular weights above 390–430 g/mol.[76] Some exceptions are known. For example, long-chain polyfluorinated compounds with molar mass >500 g/mol are taken up into plants and are translocated to leaves, though with a reduced efficiency compared to water.[77] The most important property for the accumulation of neutral and hydrophobic CECs in plants is lipophilicity, expressed as log K_{OW}; for ionizable compounds the effects caused by the dissociation of ionic molecules (e.g., ion trap effect and electrical attraction or repulsion) are often more important,[78,79] asking for a proper modeling description of their speciation in both soil and plant organs. Furthermore, many CECs (e.g., carbamazepine) are polar and reactive substances and undergo biodegradation within plants, possibly resulting in the production of more toxic by-products, but usually, after an initial oxidation step, conjugates are formed.[80,81] Apart from biodegradation, photolysis is expected to be a very effective removal process for CECs in leaves, but experimental data are scarce.

Other than CECs physical-chemical properties, which are needed for all the elements of the system, soil-crop models require an extensive amount of data to produce accurate simulations, including environmental conditions, soil and plant properties, and information on field management practices (Table 5). In case RWW reuse has already been carried out for several years,

Table 5 Data required for soil-crop modeling.

Compartment	**Data required**
Environmental conditions	– Precipitation intensity – Air temperature – Potential evapotranspiration
Soil	– Temperature – Water retention characteristics – pH – Organic carbon content – Pollutants/nutrients concentrations – Biochemical kinetic parameters incl. degradation rates
Agricultural practices	– Irrigation regime – Pollutants/nutrients concentrations in the irrigated water – Sowing and harvesting dates – Crops type, growth data and harvested amount planted

but no data on soil CECs concentrations is available, the soil concentrations can be estimated by running the coupled soil-crop model until reaching pseudo-steady-state. The soil concentrations found can then be used as initial conditions to estimate CECs uptake.

Model calibration and validation vary depending on whether the focus of the simulations is on crop fertigation or on chemical uptake due to irrigation with RWW. In the first case a comparison to reality can be carried out using growth-related data, such as grain yield, aboveground biomass and flowering and maturity days. In the second case the measured concentrations of the simulated chemicals in soil and plant organs are required. In case such concentrations are not available, a model check can be carried out using bioconcentration factors from literature. This factor is the ratio between the concentration of the chemical in various plant parts against the concentration in soil, derived from studies carried out in similar conditions. Such ratios, even though not representative of the site-specific conditions, can help to identify clear deviation between simulated uptake and experimental results.

3.3 Fate processes

In wastewater and environmental compartments, CECs can be present in multiple forms, including: (i) the dissolved fraction, both of the neutral and ionic species; (ii) the fraction adsorbed to solid matrices (sludge, sediments, soil); (iii) excreted metabolites and transformation products. The main physical-chemical and biological processes that regulate the fate of CECs throughout the entire RWW reuse system or in a specific element of the system are listed in Table 6 and described in the following sections.

Table 6 Relevant CECs processes in each element of the RWW reuse system.

	System element			
Process	**WWTP**	**Distribution system**	**Soil**	**Crops**
Sorption	x	x	x	x
Biodegradation	x	x	x	x
Deconjugation	x	x	x	
Volatilization/stripping	x	x		x
Photolysis		x	(x)	x
Abiotic hydrolysis	x	x	x	

3.3.1 Sorption

Sorption and desorption are two opposite processes that regulate the partitioning of CECs from the water phase to solid matrices, and vice versa.

The variation of CECs concentration in water with time t due to sorption and desorption can be mathematically described as:

$$\frac{dS_{CEC}}{dt} = -k_{sor}S_{CEC}X_i + k_{des}X_{CEC} \tag{1}$$

where S_{CEC} [M/L^3] and X_{CEC} [M/L^3] are the dissolved and sorbed (particulate) CECs concentrations, respectively; k_{sor} [L^3/(M·T)] and k_{des} [1/T] are the sorption and desorption rates, respectively; and X_i [M/L^3] is the concentration of the solid matrix in a defined compartment. The ratio between these two rates is equal to the solid-water distribution coefficient (K_d) [L^3/M]:

$$K_d = \frac{k_{des}}{k_{sor}} = \frac{S_{CEC}X_i}{X_{CEC}} \tag{2}$$

Sorption and desorption are assumed to be comparably fast processes, especially in wastewater solids and sludge.

Partitioning for neutral and hydrophobic chemicals is driven by the lipophilic interactions between the organic chemical and the organic matters of the solid matrix and can be described as:

$$K_d = K_{oc}f_{oc} \tag{3}$$

where K_{OC} [L^3/M] is the organic carbon-water partition coefficient, which can be expressed as function of the octanol-water partition coefficient of the chemical K_{OW},[82] and f_{OC} [M/M] is the organic carbon content of the solid matrix.

For ionic substances, sorption is less reliably described by K_{OW} since electrical forces may play a bigger role than lipophilic interactions in driving this process.[83] Hence, other relationships need to be used to estimate K_d. An example is here reported for acids and bases[84,85]:

$$K_{oc,acid} = \phi_n 10^{0.54\log P_n + 1.11} + \phi_i 10^{0.11\log P_n + 1.54} \tag{4}$$

$$K_{oc,base} = 10^{0.31\log D_{ow} + 2.78} \tag{5}$$

where ϕ_n [−] and ϕ_i [−] denote the fractions of neutral and ionized species at a reference pH_{ref} (environmental pH - 0.6), which can be derived by the Henderson-Hasselbalch relation as a function of ionization constant (pK_a)

and pH-dependent octanol-water distribution coefficient (D_{OW}). The log P_n is the log K_{OW} of the neutral molecule species (the acid).

Over time, organic compounds bound to soil (or other solid matrices) become less and less available, a process known as "aging."[86] Aging is suspected to occur by diffusion into micropores or into the solid itself, by sequestration or entrapping in organic matter, and in some cases also by labile covalent binding.[87] This process usually extends over months and years, with rates several orders of magnitude slower than those for sorption equilibria described in Eq. (2),[88] and results in the formation of a sequestered fraction that is not directly bioavailable.

3.3.2 Biotransformation

Biodegradation, also known as biotransformation, denotes the degradation of CECs into by-products by microorganisms and, less frequently, other living organisms (e.g., plants). CECs concentrations are generally too low to support microbial growth, and a number of (not mutually exclusive) mechanisms have been hypothesized or described as drivers for biodegradation, including[89–92]: (i) cometabolism or multiple substrate metabolism ("hunger metabolism"), in the presence or in the absence of one or few main growth substrates and catalyzed by non-specific enzymes, or by a diversity of enzymes; (ii) activity of highly specific enzymes and/or specialized degrader populations.

Biodegradation of CECs has mainly been observed under aerobic conditions,[93] and recent studies have shown evidence for degradation at other redox conditions (e.g., denitrifying and sulfates-reducing conditions[61,94]).

Biodegradation kinetics is most commonly described by first-order (Eq. 6) or pseudo-first-order rate equation (Eq. 7):

$$\frac{dS_{CEC}}{dt} = -k_{bio} S_{CEC} \tag{6}$$

$$\frac{dS_{CEC}}{dt} = -k'_{bio} S_{CEC} X_i \tag{7}$$

with k_{bio} [1/T] and k'_{bio} [$L^3/(M \cdot T)$] representing the biodegradation rate constants. Eq. (6) can be generally used through the whole RWW system, including WWTPs, soil and plant compartments. Eq. (7) is typically applied to model CECs fate in WWTPs, where the concentration of biomass X_i, expressed as either X_{TSS} (total suspended solids) or X_{VSS} (volatile suspended solids), can be measured or predicted efficiently by state-of-the-art process models (e.g., ASMs).

While biotransformation of CECs in WWTPs, water and soil has been widely described, less is known about metabolism-driven biotransformation in plants. When CECs are taken up into plants, they enter a living environment with numerous enzymatic processes and rapid metabolism, eventually resulting in their transformation to multiple products. According to the "green liver concept" of Sandermann,[80] metabolism in plants can be subdivided into three phases: (i) phase I is the initial oxidation reaction, which often leads to acidic or phenolic compounds; (ii) phase II is conjugation of metabolites; (iii) phase III is the compartmentalization in vacuoles or in the cell wall. Conjugation reactions, like glycosyltransferase and glutathione S-transferase, have been observed for pharmaceuticals like diclofenac, naproxen, ibuprofen, and gemfibrozil,[95,96] with glycoside and glutathione conjugates representing dominant metabolites in plant cells.

3.3.3 Deconjugation

Many CECs, especially pharmaceuticals, are present in RWW in the form of parent compounds (i.e., unchanged) and of metabolites, resulting from human metabolism and excretion after consumption. The latter fraction is rarely measured and modeled due to analytical challenges and the high numbers of metabolites for CECs. A specific type of human metabolites, i.e., conjugates, is known to retransform (deconjugate) to the respective parent compound in wastewater and environmental recipients. Although dependent on the type of conjugate,[97] this process is generally fast compared to the biotransformation of the parent compound, often leading to net formation of parent S_{CEC} within a specific compartment in the reuse system.[98] Deconjugation kinetics can be modeled similar to biotransformation by using a first-order equation for the loss of conjugates ($S_{CEC,CJ}$) and formation of parent compound (S_{CEC}):

$$\frac{dS_{CEC,CJ}}{dt} = -k_{dec} S_{CEC,CJ} \tag{8}$$

$$\frac{dS_{CEC}}{dt} = +k_{dec} S_{CEC,CJ} \tag{9}$$

with k_{dec} [1/T] representing the deconjugation rate constant.

Deconjugation has been observed in wastewater treatment systems for non-steroidal anti-inflammatory drugs (diclofenac), antiepileptic drugs (carbamazepine), antibiotics (sulfonamides) and estrogens (estriol, estradiol and estrone).[12,38,99] Furthermore, in plant metabolites, which mainly exist in the form "bound residues" after compartmentalization, can be converted back to parent compounds under specific conditions (low pH, high temperature, weathering).[81]

3.3.4 Volatilization

Volatilization is of limited relevance for most polar CECs. The mathematical formulation of volatilization kinetics can be derived from the two-film theory,[100] which leads to a first-order loss rate when atmospheric concentrations are negligible:

$$\frac{dS_{CEC}}{dt} = -k_{vol} S_{CEC} \tag{10}$$

where k_{vol} [1/T] is the first-order volatilization rate. Different empirical relationships exist in the literature to derive k_{vol} from other parameters, depending on the part of the system being modeled. For example, k_{vol} can be derived from the oxygen transfer rates in WWTP, from the environmental conditions of rivers/canals (e.g., wind velocity), from crop properties (e.g., leaf area) and intrinsic CECs properties (e.g., molar mass, air-water partition coefficient K_{aw}). The most well-known formulations are summarized in Vezzaro et al.,[101] Pomies et al.,[102] and Trapp and Harland.[100]

3.3.5 Photolysis

Photolysis is the abiotic degradation of chemicals due to the radiant energy of light either directly (direct photolysis) or due to reactions initiated by other light-absorbing compounds (indirect photolysis). The kinetics of direct photolysis are described, near the surface of water, with a first-order equation:

$$\frac{dS_{CEC}}{dt} = -k_{pho,dir} S_{CEC} \tag{11}$$

where $k_{pho,dir}$ [1/T] is the first-order photodegradation rate, which is function of other parameters such as light intensity and wavelength. Indirect photodegradation in water and air is driven by the reaction of the light-absorbing compounds leading to the formation of a variety of reactive species, like OH-radicals. For this reason, its mathematical description relies on the estimation of the concentrations of such reactive species C_{rs} [M] and their reaction rate with the target CECs $k_{pho,ind}$ [1/(M·T)]:

$$\frac{dS_{CEC}}{dt} = -k_{pho,\mathrm{ind}} C_{rs} S_{CEC} \tag{12}$$

In a simplified formulation, the near-surface photolysis rate can be used to estimate the average photolysis rate along the entire water column. For a clear mathematical description of Eqs. (11) and (12), the reader is referred to Schwarzenbach et al.[72] Both direct and indirect photolysis occur also in the soil surface layer and in crop tissues, especially leaves. These processes are affected by the complex natures of the crop tissues and the soil matrix. A comprehensive description of these processes is present in Balmer et al.[103] and Katagi.[104]

3.3.6 Hydrolysis

Hydrolysis is the reaction, through which one or multiple atoms in an organic molecule are substituted by water molecules, thus producing a more polar organic molecule. For the CECs under examination, this process is abiotic, and its kinetics is mathematically expressed as:

$$\frac{dS_{CEC}}{dt} = -k_{hyd}S_{CEC} \tag{13}$$

where k_{hyd} [1/T] is the first-order hydrolysis rate. Such value is influenced by water temperature, pH and, in case of ionic chemicals, by the speciation equilibrium. The mathematical calculation of the rate constant can be found in Schwarzenbach et al.[72] and Vezzaro et al.[101]

3.4 Available modeling tools

Several tools are available for modeling the elements of RWW reuse systems. When integrating multiple models for the single elements of a RWW reuse system, model connections and interfaces should be clearly defined. The different element models might in fact rely on different underlying assumptions, hindering a direct connection. For example, similar component might be described by different fractions: for example, organic matter can be described as biological or chemical oxygen demand; sorbed CECs can be further distinguished into fractions sorbed to particles and dissolved organic matter. Table 7 provides a small selection of the most widespread models that are available for describing nutrients and/or inorganic pollutants fluxes through a RWW reuse system. The modeler's choice would be highly case-specific, depending on factors such as simulation scope and data availability.

The following sections describe the main models available specifically describing CECs throughout a RWW reuse system.

Table 7 Scope, strengths and weaknesses of a selection of models for the simulation of RWW reuse systems, including WWTPs and surface waters.

Element	Model	Scope	Strengths	Weaknesses
WWTP	ASM family [16]	Carbon and nutrient removal	Widely used, well calibrated parameters, dynamic, implemented in several software	Organic carbon partitioning not directly comparable to laboratory results
	ASM-X [12]	Pharmaceutical removal	Dynamic, extends ASM models	Requires dedicated experimental work for calibration/validation
	IUWS_MP [105]	Addition of micropollutants to ASM	Dynamic, readily compatible with ASM models	Same as ASM models
	Cloutier et al. [11]	Heavy metals removal	Dynamic	Does not consider precipitation, lacks validation
	TOXCHEM [106]	Heavy metals and PAH removal	Considers both sorption and precipitation	Only steady-state, lacks validation
	SimpleTreat/Activity SimpleTreat [107–109]	Organic contaminant removal	Generic, publicly available, relies on easily accessible parameters	Only steady-state
Distribution system	QUAL2Kw [110]	Nutrients and organic carbon fate	Autocalibrating, detailed mass exchange between bulk-water and sediments	Only steady-state, no simulation of river branches
	WASP [111]	Nutrients, organic carbon and heavy metals fate	Dynamic, detailed heavy metals description, 2- and 3-D simulations possible	Requires extensive data, simplified sediment flux calculation
	RWQM1 [112]	Nutrients and organic carbon fate	Dynamic, readily compatible with ASM models, can be simplified to represent only selected processes/components	Not often used, requires calibration of several parameters, with identifiability problems, requires estimation of microbial biomass

	IUWS_MP [105]	Micropollutants, nutrients and organic carbon fate	Dynamic, readily compatible with ASM models	Same as RWQM1
	GREAT-ER [113]	Micropollutants fate	Use of GIS data, provides uncertainty estimates	Catchment-specific, considers only WWTP sources, only steady-state
	PhATE™ [114]	Micropollutants fate	Use of GIS data and hydrological databases, provides concentration estimates for various flow conditions	Catchment-specific, considers only WWTP sources, overly simplified degradation description
	EPANET/ EPANET-MSX [67,115]	Generic contaminants fate	Industry standard, well documented and expanded	Water quality only not along pipes, no diffusion, complete mixing at junctions
Soil and crops	CropSyst [116]	Crops growth and nutrients fate	Simulation of several crops, limited number of parameters	Calibration required, includes non-measurable parameters
	DSSAT [117]	Crops growth and nutrients fate	Simulation of several crops, several sub-models allow to describe different agronomic conditions	High number of parameters, not fully documented
	APSIM [118]	Crops growth and nutrients fate	Simulation of several crops, several sub-models allow to describe different agronomic conditions	High number of parameters, not fully documented
	WOFOST [119]	Crops growth and nutrients fate	Simulation of several crops, limited number of parameters	Simplified soil water balance, no effect of water stress on crop development
	BASL4 [120]	Micropollutants plant uptake	Few parameters required, dedicated model for carrots	Limited to crop description, simplified soil structure
	Dynamic plant uptake model [121]	Micropollutants plant uptake	Can be adapted for several crops, detailed description of ionic trapping	Includes rarely known parameters which change with time, like plant mass and soil temperatures

3.4.1 Wastewater treatment plant

Several mechanistic models to predict CECs fate in WWTPs have been developed in the past decades, both as stand-alone, *ad hoc* models, like SimpleTreat,[107,109] and as add-on of ASM models, like ASM-X[12] or IUWS_MP,[105] which includes an extension of the ASM2d. While the first is limited to steady-state conditions, the latter two can provide dynamic simulations: their range of application is therefore linked to the modeling objectives. Lautz et al.[122] showed that steady-state models can be sufficient to simulate average effluent concentrations, and can therefore be used for screening purposes. Dynamic models are often coupled with simulation of conventional pollutants, and thus allow detailed investigations of the CECs behavior within the plant (as in Baalbaki et al.[64]). This allows for investigating the effect of changing the WWTP operations (e.g., to increase the energy output of the plant, to reduce greenhouse gas emissions) on the CECs fate.

3.4.2 Distribution system

Models specifically designed to simulate the impact of CECs released from WWTPs include GREAT-ER[113] and PhATE™,[114] which use GIS data to reproduce European and North American basins, respectively. These models do not include nutrients from the RWW and diffusive (non-point source) emissions. Furthermore, due their GIS-based nature they are well suited for the simulation of whole river networks, with many point-source WWTP inputs. This is not always needed. Single river and channel stretches can be simulated with WASP[111] and IUWS_MP,[105] an extension of the RWQM1 model. In addition to the processes for simulating nutrients, these models include a detailed description of volatilization, hydrolysis, photolysis and biodegradation. IUWS_MP is implemented as a stand-alone model library (also available in commercial software platforms such as WEST®, DHI, Denmark) that allows for direct integration between the WWTP and distribution system models. Due to their level of detail, both WASP and RWQM1, require a large number of parameters.

3.4.3 Soil and crops

Several regressions and mechanistic models have been developed to model the uptake of neutral CECs from soil, such as those listed in Gobas et al.[123] While regressions are easy to use, they neglect the influence caused by many site-specific environmental conditions, like crop and soil type and uptake from air, and provide only an "average" estimation. Furthermore, as they relate solely physical-chemical characteristics of the compound to the bioconcentration factor, no information regarding the fate of CECs in soil is given. This implies the use of an additional soil model to calculate the expected concentrations in the crops. For neutral compounds, relatively simple steady-state plant uptake models are in use since decades.[68] The advantage is that those simple models can be adapted to the plant and soil under consideration, which gives usually better results.[124]

When dealing with ionic CECs, simple regressions are not able to describe the plant uptake. Two mechanistic models are available: the dynamic plant uptake (DPU) model (Trapp[121] in its latest release of January 2020), and BASL4.[120] The DPU model couples an analytical solution of a four-compartment plant model, solved via a multi-cascade approach, to a tipping bucket layered soil transport model. Partition coefficients of neutral compounds are, as usual, estimated from the log K_{OW} and the lipid and water content of the plant, while the (pseudo-)partitioning of ionizing compounds is calculated by the "cell model."[78] BASL4 is a level IV fugacity model including an equilibrium partitioning (EqP) and a dynamic carrot root (DCR) model. A comparative study conducted by Prosser et al.[125] with an earlier version of the DPU model concluded that while both models tended to overpredict uptake, due to unknown degradation rates in plants, BASL4 suffered from higher overpredictions.

4. Case studies

Two case studies providing forward predictions of CECs fate throughout a RWW reuse system are here presented. The first case study, originally presented in Polesel et al.,[126] focused on the estimation of the fate of different CECs in several typical national scenarios, starting from the emissions at household level to the accumulation in crops. In the second example, originally presented in Delli Compagni et al.,[127] models are applied to estimate the fate of CECs from a WWTP effluent, estimating ecotoxicological risk, the uptake in several crops, and the connected risk for human health, for different management scenarios.

4.1 CECs fate prediction through indirect RWW reuse and sludge application

A generic model framework was developed to describe the fate of selected CECs in an integrated system, including WWTP, surface water and soil/plant system,[126] and the associated human exposure, following ingestion of crops fertilized with sewage sludge or irrigated with freshwater receiving WWTP effluent. The approach was used to simulate a realistic worst-case scenario and included the following elements and considerations:

- Fate of CECs in WWTPs was described by the generic steady-state model Activity SimpleTreat.[108] The model represents an extension of SimpleTreat for a refined fate description of ionizable substances based on the activity concept, and it allows calculating CECs concentrations in effluent and sewage sludge.
- CECs residues from WWTPs were assumed to enter agro-ecosystems via application of biosolids and irrigation with surface water receiving WWTP effluent. Negligible CECs removal was assumed to occur during sludge stabilization. Negligible CECs attenuation and 10-fold dilution of WWTP effluent were assumed in the receiving surface water body, which was subsequently used as source of irrigation water.
- Fate of CECs in a generic agro-ecosystem was described using the dynamic plant uptake model[74] and wheat was selected as reference crop. Sewage sludge fertilization was assumed to correspond to maximum allowed rates in different EU countries, while a fixed yearly irrigation ($60\,mm\,a^{-1}$) was assumed to be provided according to national irrigation calendars (e.g., Eurostat[128]).
- The framework was adapted to different EU countries by using country specific input (substance consumption, WWTP influent concentrations) and boundary conditions (temperature, precipitation, evapotranspiration).
- A probabilistic Monte-Carlo approach was used to propagate uncertainty in model input (e.g., biodegradation rates in activated sludge and soil) to the model output.

Predicted concentrations of three CECs (triclosan, furosemide and ciprofloxacin) in wheat grain are shown in Fig. 2 for different geographical areas and reuse scenarios (sludge application, indirect RWW reuse). For triclosan and ciprofloxacin, biosolid application was considered to be the main input pathway to agro-ecosystems, due to their sorption affinity and high levels in sewage sludge. Model predictions show considerably higher uptake and accumulation for furosemide, a mobile and rather persistent compound. For the same geographical area, indirect RWW reuse (blue bar) and application of biosolids (yellow bar) showed comparable concentration levels in wheat grain. Irrigation, however, was found to result in higher relative uptake of furosemide, with bioconcentration factors being 2.5- to 4-fold higher than for application of biosolids, mostly due to the prolonged release over time and its uptake in the latter stages of the wheat growth period.

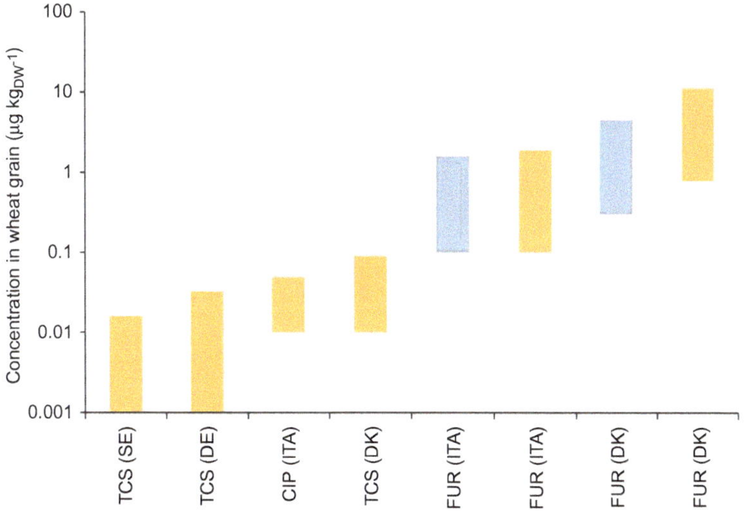

Fig. 2 Predicted concentrations of triclosan (TCS), furosemide (FUR) and ciprofloxacin (CIP) in wheat grain for different European countries (SE, Sweden; DE, Germany; ITA, Italy; DK, Denmark). Yellow bars denote biosolids application scenarios, while blue bars denote indirect RWW reuse for irrigation. Lower and upper bar boundaries denote 5th and 95th percentiles of predictions, respectively.

Differences between geographical scenarios (Italy and Denmark) could also be noticed due to country-specific furosemide human consumption. The transport of weak acids, like furosemide, in plants is strongly affected by phloem transport. It is thus worth to mention that the phloem transport path of the DPU model is still under development.

4.2 Predicting site-specific CECs consumption and health risk due to direct RWW reuse

An integrated model was developed to describe the fate of CECs in a RWW reuse system consisting of an irrigation canal for direct RWW reuse and agricultural fields, as shown in Fig. 3A.[127] A river quality model[105] and a coupled soil-plant model[121] were selected. The two original models were extended to include ionic CECs speciation and the spatial and temporal dynamics of the environmental conditions in the area.

The canal was conceptually described by using CSTR tanks in series (Fig. 3B), with each tank approximately 500 m long, and nine water withdrawal locations. Water withdrawal was calculated as the product between the water consumption to harvest one unit of land and the total irrigated area. Periods of water withdrawal were retrieved from irrigation calendars.

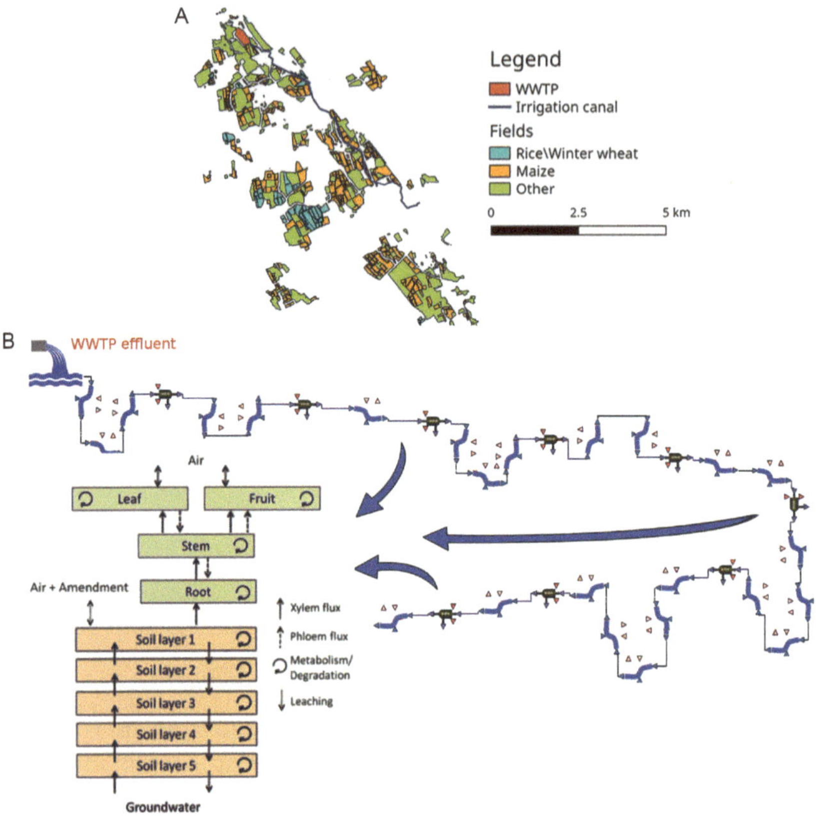

Fig. 3 (A) Outline of the RWW reuse system comprising of the WWTP, the irrigation canal and the fields of various crops. (B) System conceptualization including the canal model implemented in WEST® (DHI, Denmark) and the description of the compartments and processes included in the dynamic plant uptake model.[127]

The simulated soil layers structure was adapted to match the stratigraphy of the soils present in the area up to a depth of 1 m. Furthermore, several management scenarios, considering changes in water withdrawal, soil characteristics, crops irrigation and the presence of an upstream WWTP were tested.

Measured concentrations of conventional pollutants in the effluent were provided by the WWTP operators. Concentrations, physical-chemical properties, degradation rates and toxicity endpoints of 13 CECs in RWW were retrieved from literature and QSAR software. The physical-chemical properties of the canal water, weather data and the characteristics of various soils present in the area were collected from several public databases. Such data were either static (e.g., soil organic matter concentration) or dynamic with collection frequencies ranging from once a month (e.g., canal water temperature) to once every hour (e.g., sunlight radiation intensity). Three types of crops were simulated: rice, maize and ryegrass. Dietary consumption of the crops present in the area was retrieved from a national survey. Finally, scientific literature was also searched to select the parameters values of both the river quality and the plant uptake models and to collect crops bioconcentration factors.

The uncertainties associated to CECs transformation rates and concentrations in RWW were propagated to the predicted concentrations in the canal water. The resulting uncertainties were further propagated to the coupled soil-plant model together with crop parameter uncertainties.

The predicted canal concentrations showed adequate agreement with traditional pollutant measurements and were verified against the measurements of diclofenac (Fig. 4A—the only CECs whose measurements in the canal were available). Overall, all the simulated CECs showed limited attenuation along the canal, leading to a calculated non-negligible ecotoxicological risk in case of antibiotics (clarithromycin and sulfamethoxazole), triclosan, ibuprofen and the estrogens. The predicted bioaccumulation factors in the crop grains and the literature values used for a plausibility check are shown in Fig. 4B. Despite data are lacking for some CECs, boxplots medians were generally within 1 order of magnitude from literature values, suggesting plausible results. Even though grain accumulated CECs at concentrations that for some cases were several orders of magnitude higher than the soil, the only compounds which showed a non-negligible (but still low) risk were sulfamethoxazole and 17α-ethinylestradiol. Interestingly, this outcome is supported by Johnson et al.,[129] who found that ethinylestradiol is the CECs with the highest environmental risk. Their study was based on a comparison of available ecotoxicity data with measured concentrations in surface waters. The authors also found that, compared to the ecotoxicological risk from other chemicals, most of all heavy metals, the environmental risk from most CECs are relatively low.

Finally, scenarios considering the use of a more water-efficient irrigation system illustrated how CECs accumulation in crops could be reduced. Conversely, the use of a crop drying stage before rice harvest did not show any substantial reduction. The importance of the upstream WWTP in reducing risk was quantified by simulating a scenario where the wastewater treatment plant was neglected. In this hypothetical scenario, most CECs resulted ecotoxic and the health risk for consumers increased up to eight

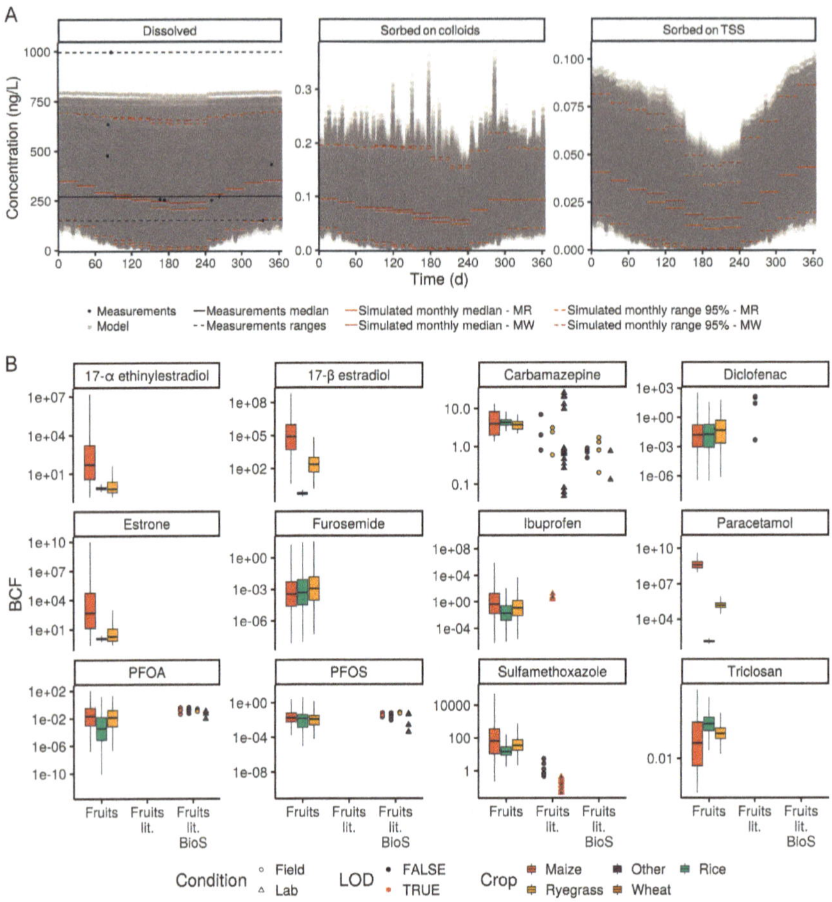

Fig. 4 (A) Diclofenac concentrations (from Jan 1st to Dec 31st) in the canal. MR (maize and rice) and MW (maize and wheat) indicate the two different water withdrawal scenarios tested. (B) Grain bioconcentration factors BCF (g_{DW}/g_{DW}) and comparison with literature values (indicated as "lit."). BioS indicates experimental studies with the use of biosolids. Red borders indicate the use of a LOD value in the calculation. See Delli Compagni et al.[127] for references.

times with respect to the current scenario with WWTP. The potential of the model for supporting the planning of future sampling campaigns was also demonstrated. For example, CECs predicted concentrations showed monthly and daily (nighttime/daytime) variations, especially in the case of diclofenac and sulfamethoxazole. This suggests that a sampling campaign performed during summer, when lower concentrations are expected, might lead to an underestimation in the CECs concentrations in the canal and irrigation water throughout the whole year.

5. Conclusions

This brief overview showed that tools for risk management in reclaimed wastewater reuse systems are available. In particular for WWTP, but also for surface waters and water distribution channels, a couple of model solutions can be chosen allowing to simulate CECs. Contrary, only few tools are available for the fate in the soil-plant system, and none of those has been rigorously tested. This pathway is also the most difficult, because transport and reaction processes in the root zone are affected by the interaction of roots, fungi, microbes, the percolating water and the soil matrix. Different from the technosphere (WWTP), these elements are not well described and uncontrolled in the rhizosphere. Fate predictions for CECs in the root zone and plant uptake are hence currently the most uncertain part of the water reuse risk management.

References

1. FAO. *Water: source of food security*. Available at: https://coin.fao.org/coin-static/cms/media/9/13171142301630/2002_26_original.pdf; 2002.
2. FAO. *Wastewater treatment and use in agriculture*. Available at: http://www.fao.org/3/T0551E/t0551e00.htm; 1992.
3. FAO. *Water quality for agriculture*. Available at: http://www.fao.org/3/T0234E/T0234E00.htm#TOC; 1994.
4. United States Environmental Protection Agency. *Aquatic life criteria for contaminants of emerging concern—part I general challenges and recommendations*. Available at: https://www.epa.gov/sites/production/files/2015-08/documents/white_paper_aquatic_life_criteria_for_contaminants_of_emerging_concern_part_i_general_challenges_and_recommendations_1.pdf; 2008.
5. Castiglioni S, Davoli E, Riva F, Palmiotto M, Camporini P, Manenti A, et al. *Mass balance of emerging contaminants in the water cycle of a highly urbanized and industrialized area of Italy*. *Water Res* 2018;**131**:287–98. https://doi.org/10.1016/j.watres.2017.12.047.
6. René PS, Escher BI, Fenner K, Hofstetter TB, Johnson CA, von Gunten U, et al. The challenge of micropollutants in aquatic systems. *Science* 2006;**313**:1072–7.
7. Kelly J, Unkovich M, Stevens D. Crop nutrition considerations in reclaimed water irrigation systems. In: Stevens D, Arris Pty Ltd, editors. *Growing crops with reclaimed wastewater*. Collingwood, Australia: CSIRO Publishing; 2006. p. 91–103.
8. WHO. *Safe use of wastewater, excreta and greywater guidelines. Volume 2: wastewater use in agriculture*. World Health Organization; 2006. https://doi.org/10.1007/s13398-014-0173-7.2.
9. European Commission. *Proposal for a regulation of the European parliament and of the council on minimum requirements for water reuse*. European Commission; 2018.
10. Singer HP, Wössner AE, McArdell CS, Fenner K. Rapid screening for exposure to "non-target" pharmaceuticals from wastewater effluents by combining HRMS-based suspect screening and exposure modeling. *Environ Sci Technol* 2016;**50**(13):6698–707. https://doi.org/10.1021/acs.est.5b03332.
11. Cloutier F, Jalby G, Lessard P, Vanrolleghem PA. *Dynamic modelling of heavy metals behavior in wastewater treatment plants*. *Rev Des Sci L'Eau* 2009;**22**:461–71. https://doi.org/10.7202/038325ar.

12. Plósz BG, Langford KH, Thomas KV. An activated sludge model for trace xenobiotic chemicals (ASM-X): assessment of diclofenac and carbamazepine. *Biotechnol Bioeng* 2012;**109**(11):2757–69.
13. Brunsch AF, Langenhoff AAM, Rijnaarts HHM, Ahring A, ter Laak TL. *In situ removal of four organic micropollutants in a small river determined by monitoring and modelling. Environ Pollut* 2019;**252**:758–66. https://doi.org/10.1016/j.envpol.2019.05.150.
14. Ort C, Hollender J, Schaerer M, Siegrist H. *Model-based evaluation of reduction strategies for micropollutants from wastewater treatment plants in complex river networks. Environ Sci Technol* 2009;**43**:3214–20. https://doi.org/10.1021/es802286v.
15. Aymerich I, Acuña V, Ort C, Rodríguez-Roda I, Corominas L. *Fate of organic microcontaminants in wastewater treatment and river systems: an uncertainty assessment in view of sampling strategy, and compound consumption rate and degradability. Water Res* 2009; **125**:152–61. https://doi.org/10.1016/j.watres.2017.08.011.
16. Henze M, Gujer W, Takashi M, van Loosdrecht M. *Activated sludge models ASM1, ASM2, ASM2d and ASM3, scientific and technical reports no. 9.* IWA Publishing; 2009.
17. Asano T, Burton FL, Laeverenz HL, Tsuchihashi R, Tchobanoglous G. *Water reuse. Issues, technologies and applications.* New York: Metcalf and Eddy, McGraw Hill; 2007.
18. Farkas J, Polesel F, Kjos M, Carvalho PA, Ciesielski T, Flores-Alsina X, et al. *Monitoring and modelling of influent patterns, phase distribution and removal of 20 elements in two primary wastewater treatment plants in Norway. Sci Total Environ* 2020;**725**:138420. https://doi.org/10.1016/j.scitotenv.2020.138420.
19. Kazadi-Mbamba C, Flores-Alsina X, Batstone DJ, Tait S. Validation of a plant-wide phosphorus modelling approach with minerals precipitation in a full-scale WWTP. *Water Res* 2016;**100**:169–83.
20. Carletti G, Fatone F, Bolzonella D, Cecchi F. Occurrence and fate of heavy metals in large wastewater treatment plants treating municipal and industrial wastewaters. *Water Sci Technol* 2008;**57**(9):1329–36.
21. Goldstone ME, Kirk PWW, Lester JN. The behavior of heavy-metals during wastewater treatment. 1. Cadmium, chromium and copper. *Sci Total Environ* 1990;**95**: 233–52.
22. Goldstone ME, Kirk PWW, Lester JN. The behavior of heavy-metals during wastewater treatment. 2. Lead, nickel and zinc. *Sci Total Environ* 1990;**95**:253–70.
23. Östman M, Lindberg RH, Fick J, Björn E, Tysklind M. Screening of biocides, metals and antibiotics in Swedish sewage sludge and wastewater. *Water Res* 2017;**115**:318–28.
24. Deavers K, Macek T, Karlson U, Trapp S. *Removal of 4-chlorobenzoic acid from spiked hydroponic solution by willow trees (Salix viminalis). Environ Sci Pollut Res* 2010; **17**:1355–61. https://doi.org/10.1007/s11356-010-0321-3.
25. ISO (International Organization for Standardization). *Water quality—fresh water algal growth test with Scenedesmus subspicatus and Raphidocelis subcapitata. ISO Standard 8692.* Geneva, Switzerland: International Organization for Standardization; 1997.
26. Urbano VR, et al. Effects of treated wastewater irrigation on soil properties and lettuce yield. *Agric Water Manag* 2017;**181**:108–15.
27. Marcar NE, Theiveyanathan T, Stevens DP. Salinity. In: Levy GJ, Fine P, Bar-Tal A, editors. *Treated wastewater in agriculture: use and impacts on the soil environment and crops.* Wiley-Blackwell; 2010. p. 286–305. https://doi.org/10.1002/9781444328561.ch8. Chapter 8.
28. Trapp S, Feificova D, Rasmussen NF, Bauer-Gottwein P. Plant uptake of NaCl in relation to enzyme kinetics and toxic effects. *Environ Exp Bot* 2008;**64**:1–7.
29. Marschner H. *Mineral nutrition of higher plants.* 2nd ed. London, UK: Academic Press; 1995.
30. Blumwald E. Sodium transport and salt tolerance in plants. *Curr Opin Cell Biol* 2000;**12**:431–4.

31. Bauer-Gottwein P, Rasmussen NF, Feificova D, Trapp S. *Phytotoxicity of salt and plant salt uptake: modeling ecohydrological feedback mechanisms. Water Resour Res* 2008;**44**(4): W04418. https://doi.org/10.1029/2007WR006067.
32. Rengasamy P. World salinization with emphasis on Australia. *J Exp Bot* 2006;**57**(5):1017–23.
33. Lieth H, Mochtchenko M, editors. *Cash crop halophytes*. Berlin: Springer; 2003.
34. Agency for Toxic Substances and Disease Registry (ATSDR). *The 2011 priority list of hazardous substances*. Online available at: http://www.atsdr.cdc.gov/SPL/index.html; 2011. Accessed 29 May 2020.
35. National Institute of Health. *National library of medicine, hazardous substances data bank (NLM/HSDB). Hazardous substances data bank*. http://toxnet.nlm.nih.gov/cgibin/sis/search; 2012. Accessed 30 May 2020).
36. European Commission Regulation. European Commission regulation (EC) No 1881/2006 of 19 December 2006 setting maximum levels for certain contaminants in foodstuffs. *Off J Eur Union* 2006;**L 364**:5–24.
37. Sauvé S, Desrosiers M. *A review of what is an emerging contaminant. Chem Cent J* 2014;**8**:1–7. https://doi.org/10.1186/1752-153X-8-15.
38. Luo Y, Guo W, Ngo HH, Nghiem LD, Hai FI, Zhang J, et al. *A review on the occurrence of micropollutants in the aquatic environment and their fate and removal during wastewater treatment. Sci Total Environ* 2014;**473–474**:619–41. https://doi.org/10.1016/j.scitotenv.2013.12.065.
39. Tran NH, Reinhard M, Gin KYH. *Occurrence and fate of emerging contaminants in municipal wastewater treatment plants from different geographical regions-a review. Water Res* 2018;**133**:182–207. https://doi.org/10.1016/j.watres.2017.12.029.
40. Scheurer M, Michel A, Brauch HJ, Ruck W, Sacher F. *Occurrence and fate of the antidiabetic drug metformin and its metabolite guanylurea in the environment and during drinking water treatment. Water Res* 2012;**46**:4790–802. https://doi.org/10.1016/j.watres.2012.06.019.
41. Janex-Habibi ML, Huyard A, Esperanza M, Bruchet A. *Reduction of endocrine disruptor emissions in the environment: the benefit of wastewater treatment. Water Res* 2009;**43**:1565–76. https://doi.org/10.1016/j.watres.2008.12.051.
42. Arvaniti O, Stasinakis A. Review on the occurrence, fate and removal of perfluorinated compounds during wastewater treatment. *Sci Total Environ* 2015;**524–525**:81–92.
43. Margot J, Rossi L, Barry DA, Holliger C. *A review of the fate of micropollutants in wastewater treatment plants. Wiley Interdiscip Rev Water* 2015;**2**:457–87. https://doi.org/10.1002/wat2.1090.
44. Verlicchi P, Al Aukidy M, Zambello E. *Occurrence of pharmaceutical compounds in urban wastewater: removal, mass load and environmental risk after a secondary treatment—a review. Sci Total Environ* 2012;**429**:123–55. https://doi.org/10.1016/j.scitotenv.2012.04.028.
45. Majewsky M, Gallé T, Bayerle M, Goel R, Fischer K, Vanrolleghem PA. *Xenobiotic removal efficiencies in wastewater treatment plants: residence time distributions as a guiding principle for sampling strategies. Water Res* 2011;**45**:6152–62. https://doi.org/10.1016/j.watres.2011.09.005.
46. Ort C, Lawrence MG, Reungoat J, Mueller JF. *Sampling for PPCPs in wastewater systems: comparison of different sampling modes and optimization strategies. Environ Sci Technol* 2010;**44**:6289–96. https://doi.org/10.1021/es100778d.
47. Verlicchi P, Ghirardini A. *Occurrence of micropollutants in wastewater and evaluation of their removal efficiency in treatment trains: the influence of the adopted sampling mode. Water (Switzerland)* 2019;**11**:1–17. https://doi.org/10.3390/w11061152.
48. Rizzo L, Gernjak W, Krzeminski P, Malato S, McArdell CS, Jose A, et al. Best available technologies and treatment trains to address current challenges in urban wastewater reuse for irrigation of crops in EU countries. *Sci Total Environ* 2020;**710**:136312.

49. Eggen RIL, Hollender J, Joss A, Schärer M, Stamm C. *Reducing the discharge of micropollutants in the aquatic environment: the benefits of upgrading wastewater treatment plants. Environ Sci Technol* 2014;**48**:7683–9. https://doi.org/10.1021/es500907n.
50. Goldstein M, Shenker M, Chefetz B. *Insights into the uptake processes of wastewater-borne pharmaceuticals by vegetables. Environ Sci Technol* 2014;**48**:5593–600. https://doi.org/10.1021/es5008615.
51. Paltiel O, Fedorova G, Tadmor G, Kleinstern G, Maor Y, Chefetz B. *Human exposure to wastewater-derived pharmaceuticals in fresh produce: a randomized controlled trial focusing on carbamazepine. Environ Sci Technol* 2016;**50**:4476–82. https://doi.org/10.1021/acs.est.5b06256.
52. Krafft MP, Riess JG. Per- and polyfluorinated substances (PFASs): environmental challenges. *Curr Opin Colloid Interface Sci* 2015;**20**:192–212.
53. Ghisi R, Vamerali T, Manzetti S. Accumulation of perfluorinated alkyl substances (PFAS) in agricultural plants: a review. *Environ Res* 2019;**169**:326–41.
54. Blaine AC, Rich CD, Sedlacko EM, Hundal LS, Kumar K, Lau C, et al. Perfluoroalkyl acid distribution in various plant compartments of edible crops grown in biosolids-amended soils. *Environ Sci Technol* 2014;**48**:7858–65.
55. WHO. *Keeping our water clean: the case of water contamination in the Veneto region. Italy.* http://www.euro.who.int/__data/assets/pdf_file/0018/340704/FINAL_pfas-report-20170530-h1200.pd; 2016.
56. Blaine AC, Rich CD, Sedlacko EM, Hyland KC, Stushnoff C, Dickenson ERV, et al. Perfluoroalkyl acid uptake in lettuce (Lactuca sativa) and strawberry (Fragaria ananassa) irrigated with reclaimed water. *Environ Sci Technol* 2014;**48**:14361–8.
57. Gredelj A, Nicoletto C, Valsecchi S, Ferrario C, Polesello S, Lava R, et al. Uptake and translocation of perfluoroalkyl acids (PFAA) in red chicory (Cichorium intybus L.) under various treatments with pre-contaminated soil and irrigation water. *Sci Total Environ* 2019;**708**:134766.
58. Rieger L, Gillot S, Langergraber G, Ohtsuki T, Shaw A, Takacs I, et al. *Guidelines for using activated sludge models.* London, UK: IWA Publishing; 2012. https://doi.org/10.2166/9781780401164.
59. Henze M, van Loosdrecht MCM, Ekama GA, Brdjanovic D. *Biological wastewater treatment. Principles, modelling and design.* London, UK: IWA Publishing; 2008.
60. Torresi E, Fowler SJ, Polesel F, Bester K, Andersen HR, Smets BF, et al. *Biofilm thickness influences biodiversity in nitrifying MBBRs—implications on micropollutant removal. Environ Sci Technol* 2016;**50**:9279–88. https://doi.org/10.1021/acs.est.6b02007.
61. Suarez S, Lema JM, Omil F. *Removal of pharmaceutical and personal care products (PPCPs) under nitrifying and denitrifying conditions. Water Res* 2010;**44**:3214–24. https://doi.org/10.1016/j.watres.2010.02.040.
62. Krzeminski P, Tomei MC, Karaolia P, Langenhoff A, Almeida CMR, Felis E, et al. *Performance of secondary wastewater treatment methods for the removal of contaminants of emerging concern implicated in crop uptake and antibiotic resistance spread: a review. Sci Total Environ* 2019;**648**:1052–81. https://doi.org/10.1016/j.scitotenv.2018.08.130.
63. Gernaey KV, Jeppsson U, Batstone DJ, Ingildsen P. *Impact of reactive settler models on simulated WWTP performance. Water Sci Technol* 2006;**53**:159–67. https://doi.org/10.2166/wst.2006.018.
64. Baalbaki Z, Torfs E, Yargeau V, Vanrolleghem PA. *Predicting the fate of micropollutants during wastewater treatment: calibration and sensitivity analysis. Sci Total Environ* 2017;**601–602**:874–85. https://doi.org/10.1016/j.scitotenv.2017.05.072.
65. Martin C, Vanrolleghem PA. *Analysing, completing, and generating influent data for WWTP modelling: a critical review. Environ Model Software* 2014;**60**:188–201. https://doi.org/10.1016/j.envsoft.2014.05.008.

66. Delli Compagni R, Polesel F, von Borries KJF, Zhang Z, Turolla A, Antonelli M, et al. *Modelling micropollutant fate in sewer systems—a new systematic approach to support conceptual model construction based on in-sewer hydraulic retention time. J Environ Manage* 2019; **246**:141–9. https://doi.org/10.1016/j.jenvman.2019.05.139.
67. Shang F, Uber JG, Rossman LA. *Modeling reaction and transport of multiple species in water distribution systems. Environ Sci Technol* 2008;**42**:808–14. https://doi.org/10.1021/es072011z.
68. Trapp S, Matthies M. *Chemodynamics and environmental modelling. An introduction.* New York, NY: Springer-Verlag; 1998.
69. Wang Y, Wang W-H, Wang R. *Simultaneous nitrification and denitrification in biofilm of a model distribution pipe fed with disinfected reclaimed water. J Water Process Eng* 2020;**35**:101207. https://doi.org/10.1016/j.jwpe.2020.101207.
70. Rossman LA, Brown RA, Singer PC, Nuckols JR. *DBP formation kinetics in a simulated distribution system. Water Res* 2001;**35**:3483–9. https://doi.org/10.1016/S0043-1354(01)00059-8.
71. Klosterman ST, Uber JG, Murray R, Boccelli DL. *Adsorption model for arsenate transport in corroded iron pipes with application to a simulated intrusion in a water distribution network. J Water Resour Plan Manag* 2014;**140**:649–57. https://doi.org/10.1061/(ASCE)WR.1943-5452.0000353.
72. Schwarzenbach RP, Gschwend PM, Imboden DM. *Environmental organic chemistry*. 2nd ed. Hoboken, NJ, USA: John Wiley & Sons, Inc; 2003. https://doi.org/10.1002/0471649643.
73. Brunetti G, Kodešová R, Šimůnek J. Modeling the translocation and transformation of chemicals in the soil-plant continuum: a dynamic plant uptake module for the HYDRUS model. *Water Resour Res* 2019;**55**(11):8967–89.
74. Rein A, Legind CN, Trapp S. New concepts for dynamic plant uptake models. *SAR QSAR Environ Res* 2011;**22**(1–2):191–215.
75. Greskowiak J, Hamann E, Burke V, Massmann G. *The uncertainty of biodegradation rate constants of emerging organic compounds in soil and groundwater—a compilation of literature values for 82 substances. Water Res* 2017;**126**:122–33. https://doi.org/10.1016/j.watres.2017.09.017.
76. Lamshoeft M, Gao Z, Resseler H, Schriever C, Sur R, Sweeney P, et al. *Evaluation of a novel test design to determine uptake of chemicals by plant roots. Sci Total Environ* 2018;**613–614**:10–9. https://doi.org/10.1016/j.scitotenv.2017.08.314.
77. Gredelj A, Polesel F, Trapp S. *Model-based analysis of the uptake of perfluoroalkyl acids (PFAAs) from soil into plants. Chemosphere* 2020;**244**:125534. https://doi.org/10.1016/j.chemosphere.2019.125534.
78. Trapp S. Plant uptake and transport models for neutral and ionic chemicals. *Environ Sci Pollut Res Int* 2004;**11**:33–9.
79. Trapp S. *Bioaccumulation of polar and ionizable compounds in plants.* In: Devillers J, editor. *Ecotoxicology modeling*. vol. 2. Boston, MA: Springer US; 2009. p. 299–353. https://doi.org/10.1007/978-1-4419-0197-2.
80. Sandermann H. Higher plant metabolism of xenobiotics: the 'green liver' concept. *Pharmacogenetics* 1994;**4**:225–41.
81. Komossa D, Langebartels C, Sandermann H. Metabolic processes for organic chemicals in plants. In: Trapp S, Mc Farlane C, editors. *Plant contamination: modeling and simulation of organic chemical processes*. Boca Raton: Lewis Pub; 1995. p. 69–106.
82. European Commission. *European Union System for the Evaluation of Substances 2.0 (EUSES 2.0). Prepared for the European Chemicals Bureau by the National Institute of Public Health and the Environment (RIVM), Bilthoven, The Netherlands (RIVM Report no. 601900005).* 2004.

83. MacKay AA, Vasudevan D. Polyfunctional ionogenic compound sorption: challenges and new approaches to advance predictive models. *Environ Sci Technol* 2012;**46**: 9209–23.
84. Franco A, Trapp S. Estimation of the soil-water partition coefficient normalized to organic carbon for ionizable organic chemicals. *Environ Toxicol Chem* 2008;**27**: 1995–2004.
85. Franco A, Struijs J, Gouin T, Price OR. *Evolution of the sewage treatment plant model SimpleTreat: applicability domain and data requirements. Integr Environ Assess Manag* 2013;**9**(4):560–8. https://doi.org/10.1002/ieam.1414.
86. Alexander M. Aging, bioavailability, and overestimation of risk from environmental pollutants. *Environ Sci Technol* 2000;**34**:4259–65.
87. Kästner M, Nowak KM, Miltner A, Trapp S, Schäffer A. Classification and modelling of non-extractable residue (NER) formation of xenobiotics in soil—a synthesis. *Crit Rev Environ Sci Technol* 2014;**44**(19):1–65.
88. Johnson MD, Keinath TM, Weber WJ. A distributed reactivity model for sorption by soils and sediments. 14. Characterization and modeling of phenanthrene desorption rates. *Environ Sci Technol* 2001;**35**:1688–95.
89. Criddle CS. The kinetics of cometabolism. *Biotechnol Bioeng* 1993;**41**(11):1048–56.
90. Egli T. How to live at very low substrate concentration. *Water Res* 2010;**44**:4826–37.
91. Fischer K, Majewsky M. Cometabolic degradation of organic wastewater micropollutants by activated sludge and sludge-inherent microorganisms. *Appl Microbiol Biotechnol* 2014;**98**:6583–97.
92. Johnson DR, Helbling DE, Lee TK, Park J, Fenner K, Kohler HPE, et al. Association of biodiversity with the rates of micropollutant biotransformations among full-scale wastewater treatment plant communities. *Appl Environ Microbiol* 2015;**81**:666–75.
93. Joss A, Zabczynski S, Göbel A, Hoffmann B, Löffler D, McArdell CS, et al. Biological degradation of pharmaceuticals in municipal wastewater treatment: proposing a classification scheme. *Water Res* 2006;**40**:1686–96.
94. Ghattas AK, Fischer F, Wick A, Ternes TA. Anaerobic biodegradation of (emerging) organic contaminants in the aquatic environment. *Water Res* 2017;**116**:268–95.
95. Bartha B, Huber C, Schröder P. *Uptake and metabolism of diclofenac in Typha latifolia—how plantscope with human pharmaceutical pollution. Plant Sci* 2014;**227**:12–20. https://doi.org/10.1016/j.plantsci.2014.06.001.
96. Wu X, Fu Q, Gan J. *Metabolism of pharmaceutical and personal care products by carrot cell cultures. Environ Pollut* 2016;**211**:141–7. https://doi.org/10.1016/j.envpol.2015.12.050.
97. D'Ascenzo G, Di Corcia A, Gentili A, Mancini R, Mastropasqua R, Nazzari M, et al. Fate of natural estrogen conjugates in municipal sewage transport and treatment facilities. *Sci Total Environ* 2003;**320**:199–209.
98. Polesel F, Andersen HR, Trapp S, Plósz B. Removal of antibiotics in biological wastewater treatment systems—a critical assessment using the activated sludge modeling framework for xenobiotics (ASM-X). *Environ Sci Technol* 2016;**50**:10316–34.
99. Göbel A, McArdell CS, Joss A, Siegrist H, Giger W. *Fate of sulfonamides, macrolides, and trimethoprim in different wastewater treatment technologies. Sci Total Environ* 2007;**372**: 361–71. https://doi.org/10.1016/j.scitotenv.2006.07.039.
100. Trapp S, Harland B. Field test of volatilization models. *Environ Sci Pollut Res* 1995;**2**(3):164–9.
101. Vezzaro L, Gevaert V, Benedetti L, De Keyser W, Verdonck F, Vanrolleghem PA, et al. *Unit process models for fate of priority pollutants. Deliverable 7.2—ScorePP project.* Technical report. 2009.
102. Pomiès M, Choubert JM, Wisniewski C, Coquery M. *Modelling of micropollutant removal in biological wastewater treatments: a review. Sci Total Environ* 2013;**443**:733–48. https://doi.org/10.1016/j.scitotenv.2012.11.037.

103. Balmer ME, Goss KU, Schwarzenbach RP. *Photolytic transformation of organic pollutants on soil surfaces—an experimental approach. Environ Sci Technol* 2000;**34**:1240–5. https://doi.org/10.1021/es990910k.
104. Katagi T. *Photodegradation of pesticides on plant and soil surfaces. Rev Environ Contam Toxicol* 2004;**182**:1–195. https://doi.org/10.1007/978-1-4419-9098-3_1.
105. Vezzaro L, Benedetti L, Gevaert V, De Keyser W, Verdonck F, De Baets B, et al. *A model library for dynamic transport and fate of micropollutants in integrated urban wastewater and stormwater systems. Environ Model Software* 2014;**53**:98–111. https://doi.org/10.1016/j.envsoft.2013.11.010.
106. Parker WJ, Monteith HD, Bell JP, Melcer H, Mac Berthouex P. Comprehensive fate model for metals in municipal wastewater treatment. *J Environ Eng* 1995;**120**:1266–83.
107. Struijs J, Stoltenkamp J, van de Meent D. *A spreadsheet-based box model to predict the fate of xenobiotics in a municipal wastewater treatment plant. Water Res* 1991;**25**:891–900. https://doi.org/10.1016/0043-1354(91)90170-U.
108. Franco A, Song L, Trapp S. *Activity SimpleTreat. User instructions.* Available at: https://homepage.env.dtu.dk/stt/Homepage%20anf/Website.htm; 2011. Accessed 29 May 2020.
109. Struijs J. *SimpleTreat 4.0: a model to predict fate and emission of chemicals in wastewater treatment plants: background report describing the equations. Bilthoven, The Netherlands.* 2014.
110. Pelletier GJ, Chapra SC, Tao H. *QUAL2Kw—a framework for modeling water quality in streams and rivers using a genetic algorithm for calibration. Environ Model Software* 2006;**21**:419–25. https://doi.org/10.1016/j.envsoft.2005.07.002.
111. Ambrose RB, Wool TA. *WASP8 stream transport-model theory and user's guide.* Washington, DC. 2017. Report from U.S. Environmental Protection Agency Office of Research and Development Washington, DC; 20460.
112. Reichert P, Vanrolleghem P. *Identifiability and uncertainty analysis of the river water quality model no. 1 (RWQM1). Water Sci Technol* 2001;**43**:329–38. https://doi.org/10.2166/wst.2001.0442.
113. Feijtel T, Boeije G, Matthies M, Young A, Morris G, Gandolfi C, et al. *Development of a geography-referenced regional exposure assessment tool for European rivers GREAT-ER. J Hazard Mater* 1998;**61**:59–65. https://doi.org/10.1016/S0304-3894(98)00108-3.
114. Anderson PD, D'Aco VJ, Shanahan P, Chapra SC, Buzby ME, Cunningham VL, et al. *Screening analysis of human pharmaceutical compounds in U.S. surface waters. Environ Sci Technol* 2004;**38**:838–49. https://doi.org/10.1021/es034430b.
115. Rossman LA. *EPANET 2. User manual. Cincinnati, OH*; 2000.
116. Stöckle CO, Donatelli M, Nelson R. *CropSyst, a cropping systems simulation model. Eur J Agron* 2003;**18**:289–307. https://doi.org/10.1016/S1161-0301(02)00109-0.
117. Jones JW, Hoogenboom G, Porter CH, Boote KJ, Batchelor WD, Hunt LA, et al. *The DSSAT cropping system model. Eur J Agron* 2003;**18**:235–65. https://doi.org/10.1016/S1161-0301(02)00107-7.
118. Holzworth DP, Huth NI, DeVoil PG, Zurcher EJ, Herrmann NI, McLean G, et al. *APSIM—evolution towards a new generation of agricultural systems simulation. Environ Model Software* 2014;**62**:327–50. https://doi.org/10.1016/j.envsoft.2014.07.009.
119. de Wit A, Boogaard H, Fumagalli D, Janssen S, Knapen R, van Kraalingen D, et al. *25 years of the WOFOST cropping systems model. Agr Syst* 2019;**168**:154–67. https://doi.org/10.1016/j.agsy.2018.06.018.
120. Hughes L, Mackay D. *Model of the fate of chemicals in sludge-amended soils with uptake in vegetation and soil-dwelling organisms. Soil Sediment Contam* 2011;**20**:938–60. https://doi.org/10.1080/15320383.2011.620046.
121. Trapp S. *Dynamic soil-plant uptake simulation model for monovalent organic chemicals, such as pesticides and pharmaceuticals. Documentation and user guide. Kongens Lyngby.* Available at: https://homepage.env.dtu.dk/stt/2017Release_Plant_Model/index.htm; 2017. Accessed 29 May 2020.

122. Lautz LS, Struijs J, Nolte TM, Breure AM, van der Grinten E, van de Meent D, et al. *Evaluation of SimpleTreat 4.0: simulations of pharmaceutical removal in wastewater treatment plant facilities*. *Chemosphere* 2017;**168**:870–6. https://doi.org/10.1016/j.chemosphere.2016.10.123.
123. Gobas FAPC, Burkhard LP, Doucette WJ, Sappington KG, Verbruggen EMJ, Hope BK, et al. *Review of existing terrestrial bioaccumulation models and terrestrial bioaccumulation modeling needs for organic chemicals*. *Integr Environ Assess Manag* 2016;**12**:123–34. https://doi.org/10.1002/ieam.1690.
124. Trapp S. Calibration of a plant uptake model with plant- and site-specific data for uptake of chlorinated organic compounds into radish. *Environ Sci Technol* 2015;**49**(1): 395–402. https://doi.org/10.1021/es503437p.
125. Prosser RS, Trapp S, Sibley PK. *Modeling uptake of selected pharmaceuticals and personal care products into food crops from biosolids-amended soil*. *Environ Sci Technol* 2014;**48**:11397–404. https://doi.org/10.1021/es503067v.
126. Polesel F, Plósz BG, Trapp S. *From consumption to harvest: environmental fate prediction of excreted ionizable trace organic chemicals*. *Water Res* 2015;**84**:85–98. https://doi.org/10.1016/j.watres.2015.06.033.
127. Delli Compagni R, Gabrielli M, Polesel F, Turolla A, Trapp S, Vezzaro L, et al. *Risk assessment of contaminants of emerging concern in the context of wastewater reuse for irrigation: an integrated modelling approach*. *Chemosphere* 2020;**242**:125185 https://doi.org/10.1016/j.chemosphere.2019.125185.
128. Eurostat. *Eurostat irrigation data*. Available at: http://epp.eurostat.ec.europa.eu; 2012. Accessed 29 January 2020.
129. Johnson AC, Donnachie RL, Sumpter JP, Jürgens MD, Moeckel C, Pereira MG. *An alternative approach to risk rank chemicals on the threat they pose to the aquatic environment*. *Sci Total Environ* 2017;**599–600**:1372–81. https://doi.org/10.1016/j.scitotenv.2017.05.039.

CHAPTER SEVEN

Feasibility evaluation in reclaimed water reuse projects through the analysis of some case studies

Paola Verlicchi* and Giacomo Zanni
Department of Engineering, University of Ferrara, Ferrara, Italy
*Corresponding author: e-mail address: paola.verlicchi@unife.it

Contents

1. Introduction 222
2. Feasibility evaluation in agricultural and recreational reuse of reclaimed water 224
   2.1 Benefits of water exchange between agriculture and city 225
   2.2 Environmental benefits evaluated by means of the shadow price method 227
   2.3 Environmental benefit evaluated by means of the hedonic price method 227
   2.4 Spatial cost benefit analysis 229
   2.5 Social benefit evaluated by means of the contingent valuation method 230
   2.6 Evaluation of agricultural, financial and environmental benefits 232
   2.7 The importance of a sensitivity analysis in CBA 234
   2.8 Environmental benefit by means of travel cost method 236
   2.9 Other methods allowing the evaluation of benefits 238
3. Decision-support models supporting reuse project evaluation 239
4. Final considerations 241

Appendix: Basic concepts in the evaluation of the economic and technical feasibility in case of reclaimed water reuse projects. 241

References 249

Abstract

In reclaimed water reuse projects, the focus is generally to develop technical issues (most adequate treatment selection, water quality in the final effluent, and satisfaction of water demand). Sometimes investments and operational and maintenance costs are also estimated with regard to the construction or upgrading of the wastewater treatment. Less frequently, projects also include an economic evaluation that considers the costs and benefits related to the project development, in order to evaluate whether the project is feasible from both a technical and an economic view point. This evaluation may be performed at different levels of analysis both in terms of costs and benefits. This chapter presents how the feasibility of reclaimed water reuse projects may be performed in both technical and economic terms through the discussion of a selection of projects for direct reuse for irrigation and recreational purposes. It emerges that

Advances in Chemical Pollution, Environmental Management and Protection, Volume 6
ISSN 2468-9289
https://doi.org/10.1016/bs.apmp.2020.07.005

different types of benefits may be considered and evaluated using different methods, including the contingent valuation method and the travel cost method. The chapter aims to provide new insights for a more comprehensive technical-economical evaluation of reclaimed water reuse projects. The approach followed is to discuss applied methodologies without entering into the analysis of the collected results in specific case studies. For this reason, data are reported only in those cases where exemplification may help understand the project development. The chapter concludes with some considerations on decision-support models: more complex tools which can help conduct a technical and economic feasibility analysis of the project.

Keywords: Agricultural reuse, Case studies, Recreational reuse, Reclaimed water, Technical and economic feasibility study

1. Introduction

The reuse of reclaimed water (RRW) is becoming an issue of increasing concern in different parts of the world and in different sectors. It is often related to irrigation demand, due to the fact that more than 70% of worldwide water consumption is for agricultural needs (https://economicsofwater.weebly.com/water-and-agriculture.html). However, reclaimed water may be directly reused for urban, industrial, recreational, and environmental purposes.[1–4]

When focusing on direct reuse for agricultural purposes, recent studies refer to new sites,[5] the risks related to specific emerging contaminants in reclaimed water (antibiotics, antibiotic resistant bacteria and antibiotic resistant genes)[6–8] microorganisms[9,10] and standards for direct reuse that is safe.[11]

The technical feasibility of reclaimed water reuse has been extensively analyzed in many investigations and much is known about the best available technologies able to produce an effluent that complies with the legal requirements for its direct reuse.[12,13] Moreover, the agronomic aspects of wastewater reuse in agriculture are well known.[6,14]

Quite often the discussions focus on technical issues, such as treatment selection and treatment reliability in fulfilling the standards for direct reuse. Less frequently, they include the analysis of investment, and the analysis of the operational and maintenance costs during the lifespan of the wastewater treatment plant, as well as a technical and economic feasibility evaluation that gives a more complete evaluation concerning technology adequateness and reliability, project costs in terms of investment, operating and maintenance, and tied benefits.

The overall economic feasibility of WWTP upgrading for direct reuse has been studied less. The common opinion is that a direct reuse for agricultural needs is possible if the farmers who could use reclaimed water can lower their expenses for irrigation if the cost of reclaimed water is cheaper than that of the conventional sources they may use: surface or groundwater.[15] For this reason, in many cases the analysis compares the costs of investments to produce an effluent adequate for a direct reuse with the cost of the water currently used.[16]

The question of irrigation expenses is certainly of primary importance, but it is not the only factor to be considered: there are many other issues which should be evaluated in order to comprehensively assess whether or not the project is feasible and in many cases they are site-specific.

A more complete economic assessment should compare, at the same time, the costs and benefits derived from wastewater treatment while simultaneously considering the total economic value of water as a productive factor (i.e., its use value) and as an environmental public good (its non-use value).

In an RRW project, the evaluation of its feasibility may be carried out at different levels and considering different parameters. They refer to economic, environmental and social spheres[17] and, depending on the study, different issues may be identified and included in the analysis.

This chapter aims to shed light on this field. A selection of case studies will be analyzed in Sections 2 and 3 and, in particular, the evaluation of the identified costs and benefits will be discussed.

The common approach to assess the economic feasibility of WWTPs for reuse is a cost-benefit analysis (CBA), which compares all the costs incurred and benefits achieved from reclaimed water reuse.

Generally, CBA is undertaken in six phases: (1) identify *all* costs and benefits, (2) value these costs and benefits in monetary units, (3) establish the time horizon of the evaluation (*n* years), (4) determine the discount rate *r*, (5) select profitability* indicators and (6) conduct a sensitivity analysis with the most uncertain variables.[18]

As many concepts and keywords referring to the feasibility studies are often used in Sections 2 and 3, they have been grouped (in alphabetic order) in Table A1 in the Appendix at the end of this chapter with a brief description or definition to help the reader become familiarized with them. Words included in this table are reported with an asterisk the first time they appear in Sections 2 and 3.

2. Feasibility evaluation in agricultural and recreational reuse of reclaimed water

The evaluation of the feasibility* of a reclaimed water reuse project for agriculture needs may be performed at various levels of complexity. Fig. 1 gives an idea of the types of parameters which could be considered in the analysis and will help the discussion which follows.

Every reuse project has specific characteristics: water quantity available for direct reuse; water quality and expected variability; WWTP reliability; distance between production point (WWTP) and users; costs of investments and exercise, and (ordinary and extraordinary) maintenance; nutrient recovery capacity; and reduction in the pollutant load released into the surface body, etc.

Reuse practice will reduce/avoid the withdrawal of fresh water sources which would then become available for other purposes, in particular, for urban area needs. From a macroscopic point of view, these can be called *interconnections* that occur between the two compartments, which in this case are agriculture and city. The interconnection concept is discussed in depth in[15] who examined the intersectoral water transfer between these two areas in an RRW. In short, the authors started from the consideration that farmers can release conventional water for cities by replacing it with reclaimed water which was originally produced by those cities or other cities. This will bring benefits to the municipalities: cost savings in terms of water extraction, water delivery, water resource development, drinking water treatment, and

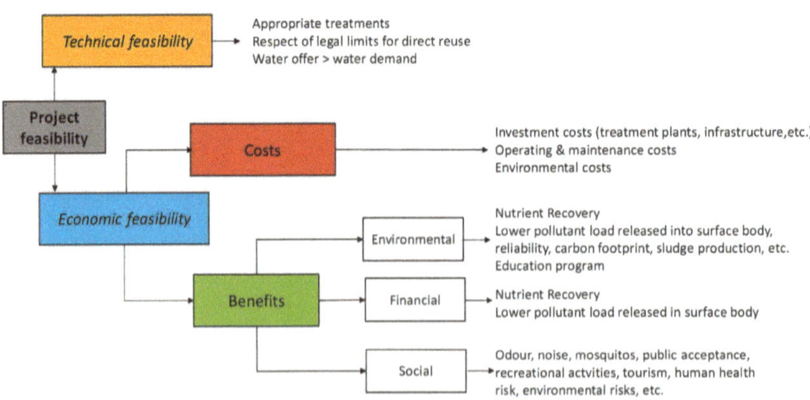

Fig. 1 Parameters which could be of interest in a technical and economic feasibility study in a reclaimed water reuse project.

removal of nutrients in wastewater treatment. In addition, there could also be economic benefits for urban areas due to the increased availability of water for industries, recreational activities and tourism, as well as benefits for the environment and, in particular, for aquatic compartments due to the reduced overexploitation of aquifers, rivers and lakes, less wastewater and pollutant loading releases and prevented seawater intrusion. From a qualitative point of view, it is quite easy to identify these issues.

Their evaluation in economic terms may be more difficult (but not impossible): tools or models could be adopted to assess their positive impacts (benefits) in monetary terms. On the other hand, as treated effluent may not (always) guarantee compliance with legal requirements (water quality standards) for its direct reuse, environmental and health risks may occur and the consequent negative impacts will decrease the benefits. The feasibility assessment of the reuse project should also consider these issues which could present difficulties in defining a monetary quantification.

2.1 Benefits of water exchange between agriculture and city

Heinz and colleagues[15] evaluated costs and benefits with regard to the reclaimed water reuse project of the Sant Feliu de Llobregat WWTP (North-East of Spain, 19 Mm^3/year).

The costs included capital costs, the operating and maintenance costs of the upgraded tertiary wastewater treatment (it was necessary to add a tertiary treatment to reduce the salt content in the WWTP effluent in order to meet the legal standards for reuse) and the cost of the infrastructure to convey the reclaimed water to the fields.

The economic benefits to the farmers included not only cost savings in water pumping (e.g., groundwater) and in fertilizing (due to the nutrient content of reclaimed water) but also increases in yields due to the constant availability of water over the whole irrigation season.

The authors remarked that the net income for farmers depends on how much they have to pay for the reclaimed water. The cost benefit analysis has to consider not only the issues related to the reuse of reclaimed water but also the fact that there will be a higher availability of water for the urban needs and the corresponding costs and benefits should be evaluated.

The analysis is thus based on this water exchange between cities and farmers who replace freshwater with treated wastewater.

The economic efficiency of the outlined water exchange depends on the costs of the withdrawal, storage and conveyance of freshwater to the cities and on the cost savings in municipal water services in the whole area of influence, corresponding to the reduced expenses in groundwater withdrawal, drinking water treatment due to the improved quality of the river water and in water resource development.

The authors completed their analysis by evaluating the economic value of this water exchange, that is of the increased municipality water supply through (Eq. 1):

$$H = FW\ x\ u \tag{1}$$

where H is the economic benefit of the water transfer, FW is the volume of freshwater released by the farmers, and u is the unitary (marginal) water value. The water value depends on the costs of water infrastructures (supply and sanitation) and can change over the time.

It may increase due to the growing cost of intersectoral water transfer; it may decrease due to the water user's reduced willingness to pay as a consequence of the increased urban water supply. In the long term, u rises with increasing water scarcity if precipitation diminishes or water demand grows.

The authors carried out an initial CBA by considering the costs of the WWTP upgrading, the construction of the pipelines to convey the reclaimed water to the users (farms) and the benefits to farmers due to the quantity of safe water always available for irrigation. Based on these initial evaluations, the project was not economically feasible.

If the CBA included intersectoral exchange, and it evaluated its corresponding costs and benefits, the projects should become viable.

This first example clearly shows that the RRW project viability* is closely related to the parameters that the CBA considers. Referring to the same case study, further environmental costs and benefits, and the corresponding impacts could also be considered. However, as is often the case, there could be many difficulties in monetarizing them.

The literature provides hints and approaches to evaluate "externalities*," i.e., the impacts exerted on non-market goods, also called "intangible" goods.[15]

Externalities can be evaluated by measuring the importance people give to the impacts concerned (such as the value of wetlands irrigated by wastewater, the creation of new green areas where people can go for relaxation, mainly in areas surrounding crowded towns). Usually, the analysis takes into consideration and also tries to quantify the *willingness* of people or communities to support the costs of obtaining the benefits or of preventing adverse impacts. If the impacts cannot be monetarized easily or not at all (e.g., certain health risks), they should be considered by using physical impact measures.

A recent attempt to evaluate health benefit is reported in[25] which makes reference to a CBA applied to the construction of a WWTP for the zootechnical wastewater in Ecuador, thus reducing the occurrence of typical diseases in the areas surrounding the farms from which the raw wastewaters are directly discharged into the surface water without any treatment. The main adverse health effects related to this practice is the spread of microorganisms in the receiving rivers which causes high numbers of gastroenteritis each year.

2.2 Environmental benefits evaluated by means of the shadow price method

A cost benefit analysis was carried out in order to evaluate the feasibility of a "centralized" desalination step for the WWTP effluent in Valencia.[26] According to this study, the proposed polishing treatment would have an environmental impact due to environmental cost savings related to not spilling (releasing) salts and an economic impact as it would guarantee an available volume of water adequate for the irrigation needs of the farmlands. The study evaluates the feasibility of the design proposal considering the internalization of environmental externalities. It was based on the shadow price*, that is a monetary value assigned to currently unknowable (or difficult-to-calculate) benefits or costs, in the absence of correct market prices (see Table A1 in the Appendix). They found that the avoided environmental costs of desalination amounted to € 62/kg salt removed. The available water volume could satisfy the demand of an area covering 272 ha for vegetable crops. The study confirmed that the adoption of the desalination stage would avoid negative impacts on soil due to a high content of salts, and at the same time an income for the farmers.

2.3 Environmental benefit evaluated by means of the hedonic price method

Apulia is a region in the South of Italy characterized by a hot and dry summer season, with a rainfall intensity of around 700 mm/year. Water for irrigation need was and is still mainly withdrawn from ground water due to the lack of surface water bodies. Around 800 million m^3 are withdrawn yearly from groundwater reserves (corresponding to 55% of the region's water demand) and 60% of this volume is used for irrigation (around 480 million m^3). Abstraction increases during drought periods. In many parts of Apulia, groundwater is the main/only freshwater source and the abstracted flow rate exceeds the recharge rate, causing water-table drawdown, well depletion, increased pumping costs, and severe seawater intrusion in coastal areas.[27] As one of the regional aims is to preserve groundwater reserves and reduce their use for irrigation, many studies have been carried out to evaluate which actions could be taken.

In this context, Arborea et al.[28] investigated two different scenarios for the irrigation reuse of reclaimed wastewater: (i) for newly irrigated land; and (ii) as a complementary source to current irrigation groundwater resources.

In brief, the authors proposed a framework for the evaluation of the costs and benefits by measuring the impacts in terms of the prices faced by the economic actor being studied and by taking a long perspective which allows for changes and upgrading of the WWTPs on a regional scale, considering the existing WWTPs.

Their analysis assumes the *WWTP gate* as the reference boundary. CBA is thus performed in terms of raw water supplies at the *plant gate* and it does not consider the costs of water conveying to the different users (farms), storing and pressurizing it, and converting it into the commodity at the place and time desired by the users.

At the same time, the benefits do not consider the value of the point of use. The benefit is associated with the steady availability of a water source (reclaimed water) over the years, which is adequate for direct reuse (in accordance with national legislation). The analysis also takes into account the economic value of guaranteeing the irrigation water supply.

As to costs, the study evaluates the investment costs (for civil works and electromechanic equipments) to upgrade (mainly rapid sand filtration followed by UV disinfection) the existing WWTPs in order to guarantee compliance with the legal limits for the direct reuse of the treated effluent, as well as the operating and maintenance costs. This estimation assumed a lifespan of 25 years for civil works and an annual maintenance rate of 1%; an annual rate of 5% for electromechanical devices with 8 years of lifespan and a straight-line depreciation rate of 4% for civil works and 12.5% for electromechanical devices.

The economic benefits are linked to the irrigation water value and to the improvement of the groundwater status (quality and quantity).

Reclaimed water reuse guarantees the steady availability of water which can increase crop yields and reduce yield variations due to drought periods (frequent in Mediterranean areas) resulting in scarce water availability for irrigation purposes.

In order to quantify this benefit, the authors followed the hedonic price method*. This method measures the implicit price of a non-market good by using the market price of a good (or service) complementary to it. For example, house prices can be used to provide the value of particular environmental attributes. Individuals may be willing to pay a premium for a house located near a country park, while they may expect a favorable price for a house located at an environmentally degraded site.

The evaluation of the economic value of good groundwater quality status, is based on the reduction of groundwater abstraction for irrigation needs and the increment in reclaimed water reuse resulting in a reduction in the salinity of the groundwater over the year up to the natural salinity value.

Arborea et al.[28] estimated the costs as a function of the WWTP size (in terms of person equivalent—PE) and of the two alternatives reported above: (i) for newly irrigated land; and (ii) as a complementary source to current irrigation groundwater resources.

To sum up, by applying the results of the evaluation by means of the hedonic price method and by assuming that there is a steady wastewater supply over the year, it was found that the average benefit derived from irrigation wastewater amounts to 0.21 €/m^3 with an annual average volume of 2475 m^3/ha.

Regarding groundwater quality status, an annual withdrawal reduction of 2080 m^3 should be applied to the level of current groundwater extraction (25,000 m^3), and it should be replaced by reclaimed wastewater. The resulting average monetary benefit is estimated at 0.22 €/m^3 with a proportional direct use value of 0.02 €/m^3 (8.3% of 0.21 €/m^3).

2.4 Spatial cost benefit analysis

In the study by Giannoccaro et al.[27] a *spatial cost-benefit analysis* of the reuse of reclaimed water was carried out referring to the same area studied in a previous one.[28]

They first evaluated the recoverable reclaimed water (by analyzing the distribution and size of the WWTPs to be included in the RRW project, their treatment trains and thus the quality of their final effluents) and the water demand for the irrigation season (around 6 months per year).

Secondly, they evaluated the economic benefits (not the environmental benefits) of the RRW. They considered the two same scenarios discussed in Arborea et al.[28] In the case of new land irrigated with reclaimed water, the costs are to upgrade the plants and the benefits include the continuous availability of water and its direct use for irrigation. In the second scenario, no additional costs are expected for reclamation and the benefit is associated with the preservation of the fresh source from salt intrusion.

The analysis showed that reclamation costs are feasible only for medium large WWTPs (greater than 5000 PE). The third step was related to the evaluation of the network necessary to distribute reclaimed water at the right places (farm gate) and, in the fourth step, the reclaimed wastewater used for irrigation was spatially located using the vulnerable groundwater map of Apulia.

The resulting data were used to assess the volume of economically feasible and readily available reclaimed urban wastewater within the groundwater vulnerable zone.

The authors found that the main technical barrier is the shortness of the irrigation season. On the other hand, the main economic concern is related to filtration followed by a lack of conveyance systems.

2.5 Social benefit evaluated by means of the contingent valuation method

In the CBA carried out by Alcon et al.[18] non-market benefits were also considered in the evaluation of the convenience of reusing reclaimed water for irrigation purposes in the South East of Spain.

This region suffers from water shortages, due to modest annual rainfall (280 mm on average), recurrent drought periods and population growth. These factors have generated an almost permanent undersupply of freshwater for the local irrigation needs.

In this context, the authors analyzed the economic and environmental implications of the reuse of reclaimed water for a mandarin farm in the Murgia Region.

Three different irrigation scenarios were considered on the farm with water from different sources and of different quality: river water (electrical conductivity (EC) ≈ 1 dS/m); mixture of surface water, groundwater and reclaimed water ($1 \leq EC \leq 3.5$ dS/m); and reclaimed water ($EC \approx 3.5$ dS/m).

The investment costs considered in this study included the costs needed to set up the plantation (soil preparation and plantation, drip irrigation technology and reservoir). The operational and maintenance costs relating to raw materials (water, energy, fertilizers and pesticides), farm labor (for irrigation maintenance, pruning, phytosanitary treatment and wood disposal) and machinery rent (for phytosanitary treatment and wood disposal). The costs of the polishing treatment of the water were not included because they are assumed to have already been paid for by the water users under the polluter pays principle.

As to the benefits, they were distinguished between private and social benefits. The private benefits depend on the obtained yields and the sale price. The yields depend on the water quality and therefore on the water source used for irrigation.

The social, non-market benefits regarded non-use values associated with preserving the ecological status of the watershed, the indirect use value associated with consumption of locally produced food, and agricultural employment. These benefits were evaluated by means of the contingent valuation method* which involved interviews with a representative sample of the population of Murgia Region.[29]

The selection of this sample must be made in compliance with the statistical representativeness, in terms of educational levels, jobs, ages, and salary, as discussed in Verlicchi et al.[30]

Respondents were first informed about the amount of money they normally paid for the treatment of wastewater and then asked if they would be willing to pay more to increase wastewater treatment up to a level that would make it suitable for direct reuse in agriculture.

They were told this would contribute to the preservation of the river's ecological status by reducing the pressure on the resource, while at the same time ensuring the water supply for agriculture.

The maximum amount the respondents were willing to pay was elicited through an open-ended question, and resulted in an average of € 5.13 per month per household on top of the current water tariff.

When adopting this value for all households in the studied area, it was found that the non-market benefits of using reclaimed water in agriculture should amount to a total annual value of € 23.3 million. This value was divided by the annual household water consumption in the area in order to obtain the benefit (€) normalized by m^3 of water consumed. It was found to be equal to 0.31 €/m^3.

The CBA completes with the calculation of the economic indicators* Net Present Value* (NPV), the Internal Rate of Return* (IRR) and the Payback Period* (PBR) (see Appendix for their definition) to evaluate the profitability of the three options.

The study by Alcon et al.[18] shows that, at the private level (i.e., for farmers), the most profitable option is mixed water, followed by surface water, while reclaimed water seems least profitable. At the social level, when environmental benefits are included, the order changes: mixed water is still the best option, but reclaimed water surpasses surface water due to its environmental benefits.

2.6 Evaluation of agricultural, financial and environmental benefits

In the study by Verlicchi et al.[31] a project of reuse (FIDEP project) was developed for the effluent of the municipal WWTP of Ferrara, Italy, to satisfy irrigation and recreational demand in the WWTP adjacent areas (agriculture needs, closed landfill cover, golf course and urban park irrigation). At the same time, the project assumed the shift of the final receiving surface water body: from the small Po di Volano canal to the River Po.

As shown in Fig. 2, the FIDEP project consisted of upgrading the current municipal WWTP (550 L/s) by means of rapid sand filtration and nature-based solutions (horizontal subsurface flow HSSF beds and lagoons) for the polishing step and the production of an effluent adequate for direct reuse.

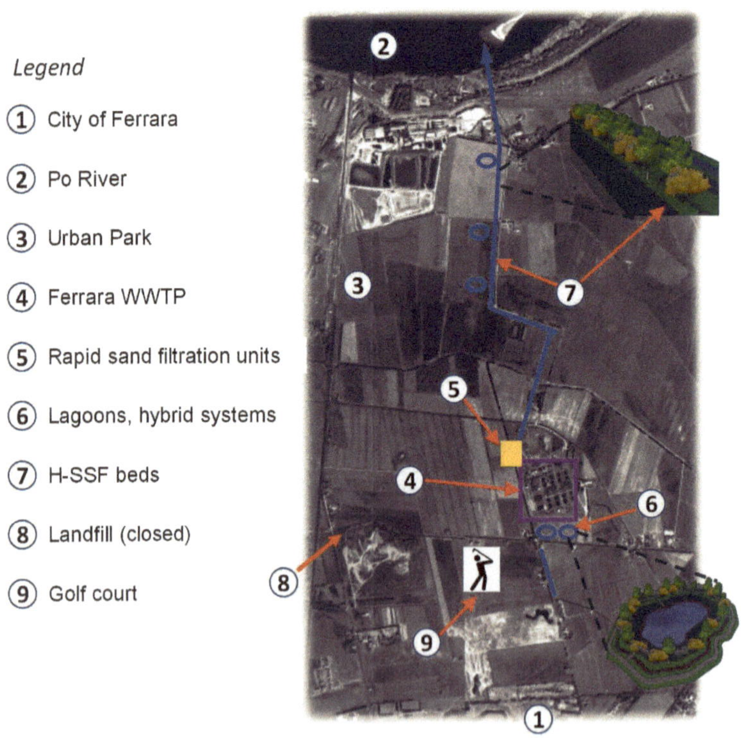

Fig. 2 Main features of the reuse project at the WWTP in Ferrara.[31]

Rapid sand filtration units were to be placed within the limits of the WWTP (number 4 in Fig. 2), whereas ponds (number 6 in Fig. 2) and HSSF beds (number 7 in Fig. 2) were placed in the park as well as in the rural area surrounding the WWTP toward the River Po (number in 3 in Fig. 2), around 3 km from the town (number 1 in Fig. 2). The HSSF beds should have a recreational value as they would also create new green areas in the park open to the public.

The estimated amount of water for direct reuse has to satisfy the request for irrigation in the rural area surrounding the WWTP, as well as the irrigation of the 19-hole golf course, the urban park, and the closed landfill cover.

The proposed solution was the result of an in-depth analysis and experimental investigations which tested different alternatives by constructed wetland pilot plants including horizontal subsurface flow systems and lagoons with particular attention to bed size and shape to avoid clogging and undesired extraordinary maintenance costs. The technical feasibility was tested by pilot plant investigations whereas the economic feasibility was based on a cost-benefit analysis.

In short, the project proposed that the whole flow rate (550 L/s) was subjected to rapid sand filtration and only the amount designated for agricultural and recreational purposes (250 L/s) had to be subjected to the further polishing treatment by means of constructed wetlands (horizontal subsurface flow beds and lagoons).

The costs included investment (the treatment plant took 2 years to be constructed, with an assumed lifespan of 20 years and a discount rate equal to 5%) as well as operating and maintenance (energy, mowing and ordinary maintenance).

As to benefits, the analysis considered: (i) the agricultural benefit derived from the reuse of reclaimed wastewater, represented by a greater and more continuous availability of water, even during periods of drought, (ii) the environmental benefit to the quality of the Po di Volano canal, due to the shift in discharge point of the final effluent from Ferrara WWTP, (iii) the financial* benefit for the Water Management Body due to the consequent reduction in energy consumption and (iv) the recreational benefit for the users of the park, resulting from the creation of green areas and ponds that are part of the polishing system.

The environmental benefit was estimated by evaluating the construction costs of an alternative project able to guarantee the same environmental benefits for the Po di Volano, i.e., the addition of a rapid filtration unit designed to treat the whole current effluent flow rate to the existing treatment plant.

The financial benefit was evaluated by considering the lower pumping costs required to move the polished effluent over the shorter distance to its new discharge point in the River Po.

As to the agricultural benefit, it was found that drought periods, which are increasingly frequent in the area of study, result in a reduction in production and consequent loss of income, up to 20–30% in some cases. An estimation of the agricultural benefit resulting from the reuse project was therefore performed by quantifying the value of the extra crop production guaranteed by the use of reclaimed wastewater.

The recreational benefit was evaluated by means of the contingent valuation method (CVM).

It consisted of providing interviewers with a questionnaire to elicit their willingness to pay* (WTP) for the construction of the polishing step at the WWTP and of constructed wetlands within the urban park fed with the treated effluent of the local WWTP.

An in-depth analysis of the questionnaire and of the sensitivity of some of the included parameters is presented in Verlicchi et al.[30] In particular, it emerged that, based on the same reference population, but on a much smaller sample (146 vs 400 interviewers), there was an 11% decrease in WTP (from € 54 to € 48.10). On the one hand, the results confirm that this method, which involves the local population, is in accordance with the "polluter pays" principle and with the recommendation to involve users in the decision-making process.

It also allows the estimation of significant externalities, which can contribute to an evaluation of the total economic value of the projects. On the other hand, the authors highlight the need to pay attention to the sampling and questionnaire design phases, in order to maintain the reliability and validity of the estimates.

FIDEP CBA was then completed by using the same economic indicators discussed above: NPV, IRR, PBP, Benefit Cost ratio* (BCR), which revealed that the proposed project was feasible. Moreover, the study highlighted that CBA is a useful tool for evaluating the suitability of a proposed project, and, combined with the treatment train selection procedure outlined above, it could prove to be a useful tool for water management authorities charged with similar decisions.

2.7 The importance of a sensitivity analysis in CBA

In the study by Galvis et al.[32] the financial viability of the reuse of treated wastewater was evaluated for sugarcane crop irrigation in the Upper Cauca river basin in Colombia.

The research included three case studies with different characteristics in terms of the existing WWTPs, WWTP size, rainfall levels and irrigation requirements (based on a water balance on the study case).

It considered two alternatives: with and without reuse. The same baseline was considered and CBA was used to compare the two scenarios.

The costs relating to the investment of upgrading the existing WWTPs and of the construction of the distribution network of reclaimed water (only in one case), as well as the operational and maintenance costs (mainly energy costs).

The benefits corresponded to avoiding the costs resulting from the implementation of wastewater reuse. They were classified into four groups: (1) savings from the reduced use of fertilizers; (2) reduction in tax for wastewater discharge directly into water bodies; (3) reduction in payment of water tariffs (freshwater use) from natural sources (groundwater); (4) savings in infrastructure (initial investment), and operation and maintenance for irrigation using groundwater.

A CBA was carried out by evaluating the benefit cost ratio BCR and the net present value NPV, assuming a time horizon of 20 years and a social discount rate equal to 11.75%. According to,[32] social discount rates are relatively high in Latin American and Caribbean countries and the assumed value is often even higher: around 12%.

The study also performed a sensitivity analysis by considering the expected variability of three main parameters: irrigation requirements, water tariff and tax for wastewater discharged on surface water bodies.

As to irrigation requirements, the three values found for the three case studies were assumed as: 0.34; 0.64; and 1.08 L/ha s. The water use fee rate varied up to 150 times the baseline value and tax for effluent released in water bodies (effluent charges) varied up to 150 times the baseline value.

The results of the CBA and sensitivity analysis showed that in the studied cases there are two key factors that influence the financial viability of treated wastewater for sugarcane crop irrigation: (1) water balance and irrigation requirements, and (2) costs corresponding to the management of wastewater for agricultural irrigation, including additional treatment (if it is required) and the infrastructure to bring the treated wastewater to crops.

As remarked above, the three case studies presented different characteristics. Only in one case was the reuse project always feasible. In another case, the infeasibility was mainly due to the low irrigation requirement and the low tariff for surface water withdrawal for irrigation.

The study makes it possible to evaluate the influence of the main parameters on the financial feasibility of the reuse projects in the different areas, which could be the most influencing parameters and if and in which conditions the project could be viable.

The authors concluded that the financial viability of reuse in the study area was limited because the values of tax for wastewater discharge and the water tariffs in Colombia do not correspond to what the values should be in order to support the financial functioning of the projects.

2.8 Environmental benefit by means of travel cost method

In the study by Segui et al.[33] reclaimed water is used for the Natural Wildlife Park of Aiguamolls de l'Emporda (Costa Brava). The authors discussed an interesting feasibility study of the wastewater reclamation and reuse project for recreational need which is based on the definition of the accurate identification of the impact of the projects, including their periodicity, quantification and evaluation. In order to do this, an optimization approach is followed, given by Eq. (2):

$$MAX(B_T) = \sum_{i=0}^{n} [(V_{ww,i} x\ SP_i) - (IC_i + OMC_i + FinC_i + T_i) + (PE_i - NE_i) - OC_i\,]_i \qquad (2)$$

where B_T is the total benefit (€); V_{ww} the annual volume of reclaimed water (m^3); *SP* the selling price of reclaimed wastewater (€/m^3); *IC* the investment costs (€/m^3); *OMC* the operational and maintenance costs (€/m^3); *FinC* the financial costs (€/m^3); T the taxes (€/m^3); *PE* the positive externalities of the impacts (€/m^3); *NE* the negative externalities of the impacts (€/m^3); *OC* the opportunity cost* (€/m^3) and *n* the years to which the figures refer.

An impact is defined as any outcome resulting from an RRW in the area of study, expected or not, promoted or accidental, and generally quantifiable. In addition, it may be internal or external. Internal impacts are those directly related to the production process of the reuse project: for instance, the income for selling reclaimed water to the different final users is a positive internal impact, while investments, operation and maintenance costs are negative internal impacts.

Given the great difficulty of identifying a standard water price* for human consumption in any country, the use of salaries was preferred. Salaries are usually accepted in empirical studies as a proxy variable that reflects the opportunity costs of taking part in this recreational activity and, therefore, have been used in this study based on the idea that people need to spend some time doing leisure activities, and this time has a value.

Impacts are generally classified into the following groups: (1) hydraulic infrastructure, (2) reuse/recovery pollutants, (3) resource use, (4) environment, and (5) education.

The study listed and quantified the impacts produced by the construction of the project in Costa Brava.

The authors found that the maximum income is mainly determined by (1) the increase in the amount of good quality water in the region, (2) the existence and preservation of the Natural Park, (3) an improvement in the quality of receiving waters, and (4) an improvement in environmental education.

In their study, the use value of the park was evaluated by means of the travel cost method*, which is an indirect economic valuation technique widely used when estimating the value of natural resources, particularly in recreational sites.[33,34] In brief, this method wants to evaluate an environmental good on the basis of the costs incurred by visitors to use the good itself. It is based on the principle that the value that an individual attributes to an environmental asset is at least equal to the costs incurred to use it. The costs include: transportation (corresponding to cost of the volume of fuel necessary to reach the place), food and accommodation; access; equipment and materials; and opportunity cost related to the visitors' time. The purpose of the procedure is to evaluate willingness to pay on the basis of the expenses incurred at the various stages of the visit (see Table A1 in the Appendix for further details of the method).

An economical evaluation of the impacts due to the reuse of pollutants (such as nutrients), or the benefits related to public health was not carried out. This was because the potential impacts are very diverse, and identification, distribution and quantification are very difficult to estimate.

Traditionally, a financial analysis of the reuse of reclaimed water focuses exclusively on the costs and private benefits. The methodology used by Segui et al.[33] considers not only the private impacts but also the project spill over that affects it. They do this through an analysis of the external impacts, both positive and negative.

The estimated use value of the reclaimed water in the Natural Wildlife Park of Aiguamolls de l'Emporda lies between 0.75, without the value of time, and 1.20 €/m^3, with the value of time.

Another interesting economic feasibility study is that reported in Alfranca et al.[35] It refers to the creation of a 1-ha wetland in the peri-urban park of Granollers, in the province of Barcelona (Catalonia), which is fed with the treated effluent of the municipal WWTP.

The study area is severely impacted by anthropic activities due to the presence of a highway, an old landfill site, a large conventional urban

WWTP (154,000 PE), a solid waste treatment plant, and a frequently used racetrack. The wetland is an elongated free water system varying between 0.3 and 1.5 m in depth, with planted (*Phragmites australis*, *Typha latifolia*) and unplanted zones.

Once the project is completed, this restored wetland will attract many people from the nearby city. The enlarged green area can thus be considered a *public good.*

The authors calculated the value of the created wetland based on the *value of water*, using the travel cost method.

In this project, the adopted methodology states the suitability of implementing a reuse project, estimates the minimum water price and maximizes the expected profit. The economic feasibility was evaluated by CBA and the economic indicator NPV.

In short, the method is generally based on the detection of the expenditure incurred by people to reach the restored area and it uses these expenditures as a means of measuring their willingness to pay for the recreational activity. The travel costs include: the estimated travel costs based on the volume of fuel used to reach the wetland; the recreational costs arising from the need to have some food during the journey or when reaching the green area, or from spending several nights in accommodation either during the journey or at the recreational site; and the opportunity costs of time spent (measured on the basis of wages).

In Ref. 35, the travel costs were not considered as it was assumed that visitors are generally inhabitants of the nearby city and that they would reach the green area by bike or on foot. The recreational costs were not considered either given the local nature of the urban park.

The opportunity cost evaluation was based on the minimum interprofessional wage for Spain. As this study was conducted in 2009, it was equal to € 18.03/day (€ 2.25/h). Assuming that a visitor spends 1.5 h in the park, including travel to reach the site, the annual opportunity cost will be evaluated on the basis of the total number of visitors in a year.

The authors outlined that not all benefits were considered, for instance, the profitability of the system would increase if the reuse of nutrients was included in the evaluation.

2.9 Other methods allowing the evaluation of benefits

In Table A1, other methods are briefly reported for the evaluation of benefits. These include the contingent choice method, which is quite similar to the contingent evaluation method and where the respondent has to choose from the alternatives provided, and the value transfer method*, which consists of the evaluation of an environmental asset in one area (policy site) on the basis of estimated values for the same environmental asset in another area (study site).

3. Decision-support models supporting reuse project evaluation

Another approach in evaluating the feasibility of the reclaimed water reuse project is the adoption of optimization models. They may be single-objective[36] or multi-objective.[17] They may include economic, environmental and social indicators[36,37] and assume location, technology, capacity and flow as decision variables.

A rapid look at some of these models highlights their strengths and complexity.

The study by Rezaei et al.[37] developed a multi-objective optimization model in order to minimize the costs and carbon footprint (selected as the environmental indicator), and to maximize the value of resource recovery (representing a social benefit) of the water reclamation system, by locating the treatment facility, allocating the treatment capacity, selecting the treatment technology, and allocating customers (final reclaimed water users). An example is also provided of its application: the expansion of the water reclamation system in Hillsborough County, Florida. From an economic point of view, it was found that a centralized plant was the most convenient solution. However, if environmental indicators are considered, the solution with a decentralized system was the most feasible option. This was mainly due to the relevant decrement in water transfer needs in less densely populated areas. A sensitivity analysis was also performed to provide insight into the main factors influencing the optimal configuration of water reclamation systems. It was found that sustainable solutions to the management of wastewater systems (e.g., decisions pertaining to the effluent water quality, location of facilities, and the decentralization of treatment facilities) highly depends on case specific parameters such as the population density and topography of the water service area. This study highlights the need for comprehensive sustainability assessments that simultaneously account for *economic, environmental, and social dimensions*, as well as the usefulness of multi-objective optimization tools in this regard that allow the decision-makers and stakeholders to weigh up the trade-offs against the objectives and design options.

Reclaimed water can be used in different applications, each characterized by different economic, environmental and social issues. Decision-makers may be supported in evaluating the feasibility of different scenarios by adopting a multi-criteria framework. In this context, the framework proposed by Chhipi-Shrestha et al.[38] combined a multi-criteria analysis* and game theory for the selection of a sustainable water reuse project. The evaluation was based on environmental criteria or indicators (fresh water saving, energy use, and carbon emissions), economic criteria (annualized life cycle cost); and social criteria (government policy, public perception); and human health risk for three stakeholders (municipality, citizens, and farm operators). The game theory was applied to eight water reuse options. The result showed that non-potable water reuse for lawns, golf course and public park irrigation, and toilet flushing is the best solution, with the mutual sharing of benefits between the municipality and citizens. The scenario analysis showed that the weights of sustainability criteria are important in decision-making. Also, the sensitivity analysis shows that the change in the amount of reclaimed water availability can affect water reuse sustainability performance.

Other tools of different complexity have been developed in order to assist and support decision-makers in the selection of treatments and the size of the plants. For instance, Verlicchi et al.[39] proposed an index, called wastewater polishing index (WWPI), which compares the water quality level achieved by different polishing treatment trains, on the basis of a set of parameters (BOD_5, COD, TSS, total P, NH_4, and *Escherichia coli*), and its proximity to law thresholds for reuse. Hidalgo et al.[40] proposed a multi-criteria software with a rough evaluation of unit process design, alternative schemes, and possible final uses. Various issues were considered in the evaluation including land availability, type of soil, type of crops, water requirement, and meteorological conditions. Papa et al.[41] proposed a decision support system (DSS) for the evaluation of the feasibility of reclaimed water reuse projects, based on three main parameters (so called "actors"): the WWTP; the hydraulic system (HS), required to transport water from the plant gate to the user; and the final user (U) (e.g., crops irrigation). A list of input parameters was given for each actor. They were selected in order to clearly define each actor (for instance: concentrations of the main parameters at the inlet and outlet of the WWTP; investment costs of the hydraulic system; water requirement for crops; and maximum concentration of pollutants in the reused effluent for the final user in the case of agriculture reuse). Then an index score has to be assigned to each input factor by comparing its measurement with target values (typically, in the case of WWTP, the standard for reuse or discharge), by means of normalization/rating curves. The resulting indices, optionally weighted, are combined into a final indicator for each actor: F_{WWTP}, F_{HS}, and F_{U}. The last step is the evaluation of the average value F of these three final indicators. According to the choice made by the authors, F will vary from +1 (best performance) to 1 (worst performance). This evaluation tool allows a rapid and prompt answer for decision-makers when comparing different reuse projects on the basis of a detailed lists of parameters that characterize the various cases to be compared.

4. Final considerations

The discussion reported herein highlighted that reuse project feasibility may be performed considering different levels of complexity and parameters among economic, social and environmental fields. The results may change according to their selection. Moreover, a sensitivity analysis may help evaluate how the results may change according to variations in the assumptions made in the baseline scenario and alternatives, and the corresponding economic and technical feasibility.

Decision-support models can help evaluated whether a project is viable, however, even for these models, the level of complexity may change and an in-depth analysis of their characteristics may result in identifying the best tool for each specific case.

Appendix: Basic concepts in the evaluation of the economic and technical feasibility in case of reclaimed water reuse projects.

See Table A1.

Table A1 Basic concepts in the economic and technical feasibility studies of reclaimed water reuse projects.[19–24]

Concept	Definition or description
Contingent choice method	Contingent choice is a technique similar to contingent valuation. However, instead of asking the respondents to directly express the value for an environmental good in the questionnaire, they are asked to express their preferences by choosing between different alternatives. Each alternative is characterized by different values of one group

Continued

Table A1 Basic concepts in the economic and technical feasibility studies of reclaimed water reuse projects.[19–24]—cont'd

Concept	Definition or description
	of attributes (environmental characteristics or services) that imply a different price or cost to the individual. Considering that each group of attributes is associated with a monetary value (for example, a payment), it is possible to reconstruct the monetary value of the environmental asset, based on the choices made by the interviewees. Compared to the contingent valuation, this method is more robust, as the choice between different alternatives facilitates the interviewee's expression of preferences. It is particularly suitable for policy decisions, in which a set of possible actions can result in different impacts on environmental assets, as it pushes the interviewee to find the best trade-off among options with different characteristics. For this reason, it is extremely useful in the evaluation of environmental improvement measures, where several flows of effects and services are affected simultaneously. For example, improved water quality in a reservoir could improve the quality of several services provided (drinking water, fishing, recreation and biodiversity, etc.). The results of this method can assign a monetary value to the environmental resource (as in the case of the contingent valuation), but can also be used to draw up a ranking of different policy options
Contingent valuation	The contingent valuation (CV) consists of the creation of a hypothetical market in which the population's willingness to pay for an environmental good is detected. The survey takes place through a questionnaire, in which a sample of interviewees is asked to attribute a value to the asset being assessed. CV is now accepted as an evaluation method in many countries, despite the limitations of robustness and the quality of the estimate when carried not carried out correctly. Following an oil pollution event, the National Oceanic and Atmospheric Administration (NOAA) panel[19] developed guidelines on how to make assessments with the CV method more reliable. The guidelines focus on the following phases: identification of the reference population and the methods of sampling and interview; preparation of the questionnaire; administration of the questionnaire; data processing and value calculation

Table A1 Basic concepts in the economic and technical feasibility studies of reclaimed water reuse projects.[19–24]—cont'd

Concept	Definition or description
Cost benefit analysis (CBA)	Cost benefit analysis (CBA) is an economic technique that assesses a government project (or program) by determining whether societal welfare will increase because of it. CBA first received official government recognition under the U.S. Flood Control Act of 1936 The aim was to evaluate the benefits and costs of water resource projects, in the interests of social welfare Today, CBA is broadly used to evaluate choices between project alternatives in a wide range of areas including natural resources, agriculture, health, transportation and business CBA assigns a monetary value to each input and each output resulting from the project, for each year of its life cycle and compares the full costs with the value of the benefits. To take into account the social preference of current consumption over future consumption, the value of benefits and costs is reduced as they move away over time. The process of reducing the value of benefits or costs is called *discounting*. This is done by multiplying the costs or benefits by a fractional number depending on the social discount rate (r) and on how far in the future they occur (k). The analyst can then calculate the net benefits of the project, examine the ratio of benefits to costs, determine the rate of return on the original investment, and compare the project results with those of other alternatives The CBA carries out the evaluation through some economic indicators: net present value, benefit cost ratio, internal rate of return and payback period, which are defined below
Cost of water and price of water	It is necessary to distinguish between the cost of water, which corresponds to the real cost of its management and distribution for the firm which "produces" it, and the price of water, which is the part of this cost invoiced to users. Sometimes, the latter is sometimes higher than the former, other times lower and insufficient to ensure the quality, maintenance and renewal of the facilities. The cost of water comprises construction costs and provisions for the replacement of all water collection; treatment and distribution facilities; cost of treatment, operation and maintenance, including that of the pipe network;

Continued

Table A1 Basic concepts in the economic and technical feasibility studies of reclaimed water reuse projects.[19–24]—cont'd

Concept	Definition or description
	administrative and customer management services; and resource preservation costs. In addition to all this, taxes of various kinds are often charged by local or state authorities. Where there are sanitation and wastewater treatment services, the costs of these services are also added. There are two policy lines regarding the question of what cost the user has to pay: that of "full cost recovery" and that of "sustainable cost recovery." Full cost recovery consists of charging users for the total cost of water. This policy aims to ensure sufficient cash flow to allow the agency managing water supplies to function sustainably in the long term. However, this can lead to failures, if the water price to be paid to users is prohibitive. The "sustainable cost recovery" model has imposed itself in most cases. Opinions on which components are to be covered by users vary from case to case. In general, there is consensus that the price to be paid should at least be equal to the operating, treatment, and maintenance costs, while construction and amortization expenses, should be covered by external financing, from the state or international bodies
Economic indicators	*Net Present Value (NPV)* is defined as the difference between the total discounted benefits and the total discounted costs, as shown in Eq. (A1), where C_k and B_k are the costs and the benefits, respectively, at year k, r the discount rate and n the expected project life cycle $$NPV = \sum_{k=0}^{n} \frac{-C_k + B_k}{(1+r)^k} \quad \text{(A1)}$$ If $NPV \geq 0$ the implementation of the project is acceptable from an economic point of view, if $NPV < 0$ the investment should be rejected as the costs are greater than the benefits *Benefit Cost Ratio (BCR)* is defined as the present value of benefits divided by the present value of costs (Eq. A2): $$BCR = \frac{\sum_{k=0}^{n} \frac{B_k}{(1+r)^k}}{\sum_{k=0}^{n} \frac{C_k}{(1+r)^k}} \quad \text{(A2)}$$

Table A1 Basic concepts in the economic and technical feasibility studies of reclaimed water reuse projects.[19–24]—cont'd

Concept	Definition or description
	Internal Rate of Return (IRR) is defined as the discount rate that equates the present benefit value and the present cost value *Pay Back Period (PBP)* is defined as the first year in which the benefit accumulation covers the cost accumulation The implementation of the project is acceptable if $BCR \geq 0$, $PBP \leq n$ and $IRR \geq r$ and if $\text{PBP} \leq n$
Externalities	An externality is any cost or benefit that spills over from the project toward other parties without any monetary compensation. Environmental effects are typical externalities in the context of CBA. Due to their nature, externalities are not included in the evaluation of the direct costs and benefits of the project. Therefore, they need to be evaluated separately and added to the count. A willingness to pay (WTP) approach should be adopted to include these effects into the appraisal. In principle, the WTP estimated for the project should capture these external effects. Examples of areas of non-market impacts of a project are changes in: life expectancy or quality of life; fatalities; injuries or accidents; landscape quality; noise; resilience to current and future climate change; travel time; and vulnerability and risk
Feasibility	In general terms, feasibility is the possibility that something can be made, done or achieved. In CBA, feasibility is the situation in which the advantages of a project are greater than the disadvantages. Feasibility can be intended in a financial sense (when the inflows are higher than the outflows) or in an economic sense (when the economic benefits are higher than the economic costs). Examples: "*We're looking at the (financial) feasibility of a new business*"; "*The state commissioned a report on the (economic) feasibility of a more advanced health system*"
Financial and economic analysis	When the feasibility of a public interest project is verified with the CBA, two elaborations are usually carried out jointly. The first (financial analysis) is set from the manager's point of view (for example, the agency called to manage the project), as if it were a business. The aim is to verify whether the outflows and cash inflows of the project make long-term management sustainable

Continued

Table A1 Basic concepts in the economic and technical feasibility studies of reclaimed water reuse projects.[19–24]—cont'd

Concept	Definition or description
	The second (economic analysis) represents the public assessment and is conducted taking into account the interest of the company as a whole. The aim is to compare the benefits with the social costs of the project, and check whether it meets the goal of achieving a net improvement in social welfare
Hedonic price method	The hedonic price consists of the evaluation of a good (for which there is no market price), on the basis of other private goods (for which there is a market price) whose value is, in some way, a function of the subject good. The connection between the two goods is typically constitutes a complementary relationship, which has a mutual influence on the value. The application often concerns real estate and environmental assets. For example, it is possible to estimate the value of the quality of water resources on the basis of the influence that the state of the water of a basin exerts on the value of nearby real estate, compared to more distant properties. The method can be applied through the following phases: identification of the available information bases; statistical analysis to identify the contribution of the environmental good to the value of market goods; and calculation of the value of the environmental asset
Multi-criteria analysis	Multi-criteria analysis is a set of methods aimed to choose between multiple alternatives on the basis of multiple evaluation criteria. They find wide application in the environmental sector as the effects to be taken into account in public and private choices are particularly varied and lend themselves to being evaluated using multiple dimensions. Multi-criteria decision analysis is divided into two categories: multi-attribute analysis, which concentrates on problems with discrete decision spaces; and multi-objective analysis, which involves several competing objectives that are required to be optimized simultaneously. From a practical point of view, the former is applied to problems where the number of alternatives is predetermined: the decision maker is to evaluate a finite number of options. The second is applied to problems in which the options have been non-predetermined. In multi-objective problems, the set of

Table A1 Basic concepts in the economic and technical feasibility studies of reclaimed water reuse projects.[19–24]—cont'd

Concept	Definition or description
	admissible solutions can be represented by a continuous space, while in multi-attribute problems they can be represented in a set of finite points. The different form of the "feasible-set" has conditioned the development of algorithms aimed at solving decision problems. In multi-objectives, the most important aspect is the representation of the objective function, resorting to more or less sophisticated mathematical programming techniques. In multi-attribute, since the set of eligible solutions is explicit and complete, the interest is focused on the formulation of algorithms that simulate the choice processes
Opportunity cost	The concept of opportunity cost is used in economics to value resources or assets. It is distinct from production cost, which consists of the sum of the costs incurred to acquire them. The opportunity cost is measured by reference to the opportunities foregone at the time an asset or resource is used. For example, the use of labor in a voluntary activity can be valued at the wages which could have been earned in a paid job. Similarly, the change in the amount of water in an irrigation system can be estimated on the value that would have been obtained (or lost) thanks to its use on crops
Profitability	In general terms, profitability is the degree in which a company, or a single business, produces a profit (money earned above what it costs to produce and sell goods and services). In CBA, profitability is mainly used in the financial analysis of projects. Example: "*The general manager is satisfied with the expected level of profitability of this business*"
Shadow prices	When we conduct a financial analysis, market prices represent a relevant signal for the private (or the public) investor assessing the project's financial performance. However, they are no longer relevant when we do an economic analysis, with the aim of assessing the project contribution to economic welfare. For this purpose, all revenues and costs considered to analyze the financial performance and valued at prices observed on the market, have to be valued adopting "shadow prices." A shadow price is the social marginal value of an output or input change, that is to say the opportunity cost to society of producing or consuming more or less of any good

Continued

Table A1 Basic concepts in the economic and technical feasibility studies of reclaimed water reuse projects.[19–24]—cont'd

Concept	Definition or description
Travel cost method	The travel cost method is used to estimate the economic value of environmental goods that are suitable for visits, such as parks or recreational sites. It can be used to evaluate the benefits or costs resulting from: elimination of an existing site; addition of a new one; changes in access or environmental quality. The method consists of evaluating an environmental good on the basis of the costs incurred by visitors to use the good itself. It is based on the principle that the value that an individual attributes to an environmental asset is at least equal to the costs incurred to use it. The costs include: transportation (corresponding to cost of the volume of fuel necessary to reach the place), food and accommodation; access; equipment and materials; and opportunity cost related to the visitors' time. The purpose of the procedure is to evaluate willingness to pay on the basis of the expenses incurred at the various stages of the visit. For this purpose, the data collected through the interviews are used to reconstruct a demand function of the environmental good, in relation to the distance and characteristics of the potential visitor population. This is analogous to estimating peoples' willingness to pay for a marketed good based on the quantity demanded at different prices. The area underlying the demand function represents the value of the environmental good; the changes in the area underlying the demand function represent the value of the change in the quality (or quantity) of the good
Value transfer	The value transfer (or benefit transfer) consists of the evaluation of an environmental asset in one area (policy site) on the basis of estimated values for the same environmental asset in another area (study site). The value transfer can be achieved in three different ways: (i) unadjusted value transfer, in which the values used for the policy site are the same (not adapted) values recorded at the study site; (ii) adjusted value transfer, in which the values are adapted on the basis of significant differences in the variables that condition the alleged environmental value and which are different at the policy site compared to the study site; (iii) transfer of the benefit function, in which it is used the function that binds the value to the

Table A1 Basic concepts in the economic and technical feasibility studies of reclaimed water reuse projects.[19–24]—cont'd

Concept	Definition or description
	explanatory variables, estimated at the study site to evaluate the value of the asset at the policy site, inserting the values of the explanatory variables relating to the policy site in the function. These three ways are sorted by increasing accuracy. Therefore, the use of the third option is recommended, when the necessary information is available
Viability	In general terms, viability is the level of chance that something will work as intended (or will succeed). In CBA, viability is sometimes used as a synonym for feasibility. Example: "It is necessary to assess the viability of all investment options under different scenarios"
Willingness to pay (WTP)	The WTP measures the maximum amount people would be willing to pay to gain outcomes that they view as desirable or, alternatively, the maximum amount that people would be willing to pay to avoid outcomes they view as undesirable. The project outputs may include both goods and services currently sold on the market and externalities. Even if a tariff is paid by users, this could be distorted and not reflect the total cost of production or the additional social benefit and cost entailed in the production of that good/service. In such situations, the correct socioeconomic value of the output is not the observed tariff, but the WTP. The importance of using the WTP is evident also if externalities are produced by the project. The externalities have to be valued in monetary terms and considered in the economic analysis of the project. The WTP provides the total welfare improvement, taking into account welfare changes in all gainers and losers of the project

References

1. Cherchi C, Kesaano M, Badruzzaman M, Schwab K, Jacangelo JG. *Municipal reclaimed water for multi-purpose applications in the power sector: a review. J Environ Manage* 2019;**236**:561–70. Available from https://doi.org/10.1016/j.jenvman.2018.10.102.
2. Kellis M, Kalavrouziotis IK, Gikas P. Review of wastewater reuse in the Mediterranean countries, focusing on regulations and policies for municipal and industrial applications. *Global NEST J* 2013;**15**(3):333–50.

3. Ortega E, Iglesias R. Reuse of treated municipal wastewater effluents in Spain: regulations and most common technologies, including extensive treatments. *Desalin Water Treat* 2009;**4**(1–3):148–60.
4. Chhipi-Shrestha G, Hewage K, Sadiq R. *Fit-for-purpose wastewater treatment: conceptualization to development of decision support tool (I). Sci Total Environ* 2017;**607–608**:600–12. Available from https://doi.org/10.1016/j.scitotenv.2017.06.269.
5. De Las HJ, Mañas P. Reclaimed wastewater to irrigate olive groves and vineyards: effects on soil properties. *Agronomy* 2020;**10**(5):1–15.
6. Christou A, Agüera A, Bayona JM, Cytryn E, Fotopoulos V, Lambropoulou D, et al. The potential implications of reclaimed wastewater reuse for irrigation on the agricultural environment: the knowns and unknowns of the fate of antibiotics and antibiotic resistant bacteria and resistance genes—a review. *Water Res* 2017;**123**:448–67.
7. Polesel F, Plósz BG, Trapp S. From consumption to harvest: environmental fate prediction of excreted ionizable trace organic chemicals. *Water Res* 2015;**84**:85–98.
8. Delli Compagni R, Gabrielli M, Polesel F, Turolla A, Trapp S, Vezzaro L, et al. *Risk assessment of contaminants of emerging concern in the context of wastewater reuse for irrigation: an integrated modelling approach. Chemosphere* 2020;**242**:125185. Available from https://doi.org/10.1016/j.chemosphere.2019.125185.
9. Caicedo C, Rosenwinkel KH, Exner M, Verstraete W, Suchenwirth R, Hartemann P, et al. *Legionella occurrence in municipal and industrial wastewater treatment plants and risks of reclaimed wastewater reuse: review. Water Res* 2019;**149**:21–34. Available from https://doi.org/10.1016/j.watres.2018.10.080.
10. Ternes TA, Prasse C, Eversloh CL, Knopp G, Cornel P, Schulte-Oehlmann U, et al. Integrated evaluation concept to assess the efficacy of advanced wastewater treatment processes for the elimination of micropollutants and pathogens. *Environ Sci Technol* 2017;**51**(1):308–19.
11. Rock CM, Brassill N, Dery JL, Carr D, McLain JE, Bright KR, et al. Review of water quality criteria for water reuse and risk-based implications for irrigated produce under the FDA food safety modernization act, produce safety rule. *Environ Res* 2019;**172**:616–29.
12. Rizzo L, Gernjak W, Krzeminski P, Malato S, McArdell CS, Perez JAS, et al. *Best available technologies and treatment trains to address current challenges in urban wastewater reuse for irrigation of crops in EU countries. Sci Total Environ* 2020;**710**:136312. Available from https://doi.org/10.1016/j.scitotenv.2019.136312.
13. Deviller G, Lundy L, Fatta-Kassinos D. Recommendations to derive quality standards for chemical pollutants in reclaimed water intended for reuse in agricultural irrigation. *Chemosphere* 2020;**240**:124911.
14. Gatta G, Libutti A, Gagliardi A, Beneduce L, Brusetti L, Borruso L, et al. *Treated agro-industrial wastewater irrigation of tomato crop: effects on qualitative/quantitative characteristics of production and microbiological properties of the soil. Agric Water Manag* 2015;**149**:33–43. Available from https://doi.org/10.1016/j.agwat.2014.10.016.
15. Heinz I, Salgot M, Koo-oshima S. Water reclamation and intersectoral water transfer between agriculture and cities—a FAO economic wastewater study. *Water Sci Technol* 2011;**63**(5):1067–74.
16. Nurizzo C, Bonomo L, Malpei F. Some economic considerations on wastewater reclamation for irrigation, with reference to the Italian situation. *Water Sci Technol* 2001;**43**(10):75–81.
17. Molinos-Senante M, Gómez T, Caballero R, Hernández-Sancho F, Sala-Garrido R. *Assessment of wastewater treatment alternatives for small communities: an analytic network process approach. Sci Total Environ* 2015;**532**:676–87. Available from https://doi.org/10.1016/j.scitotenv.2015.06.059.
18. Alcon F, Martin-Ortega J, Pedrero F, Alarcon JJ, de Miguel MD. Incorporating non-market benefits of reclaimed water into cost-benefit analysis: a case study of irrigated mandarin crops in southern Spain. *Water Resour Manag* 2013;**27**(6):1809–20.

19. Arrow K, Solow R, Portney PR, Leamer EE, Radner R, Schuman H. Report of the NOAA panel on contingent valuation. *Fed Regist* 1993;**58**(10):4601–14.
20. Atkinson G, Mourato S, Pearce DW. *Cost-benefit analysis and the environment. Recent developments*. Paris: OECD Publishing; 2006.
21. Boardman AE, Greenberg DH, Vining AR, Weimer DL. *Cost-benefit analysis: concepts and practice*. 3rd ed. Upper Saddle River, New Jersey: Pearson Prentice Hall; 2006.
22. European Commission Guide to Cost-Benefit Analysis of Investment Projects. *Economic appraisal tool for Cohesion Policy 2014–2020*. ISBN 9789279347962. (2014).
23. Pearce D. *Cost-benefit analysis*. 2nd ed. Basingstoke: Macmillan; 1983.
24. Pearce DW, Atkinson G, Mourato S. *Cost-benefit analysis and environment: recent developments*. Paris: OECD; 2006.
25. Verlicchi P, Estefanía Caridad AS, Zanni G. Zootechnical farm wastewaters in Ecuador: a treatment proposal and cost-benefit analysis. *Water (Switzerland)* 2019;**11**(4): 1–18.
26. Bellver-Domingo A, Fuentes R, Hernández-Sancho F. Reclaimed water and irrigation: a cost-benefit analysis for desalination of WWTPs effluents. *Desalin Water Treat* 2019;**166**:193–201.
27. Giannoccaro G, Arborea S, de Gennaro BC, Iacobellis V, Piccinni AF. Assessing reclaimed urban wastewater for reuse in agriculture: technical and economic concerns for Mediterranean regions. *Water (Switzerland)* 2019;**11**(7):1511.
28. Arborea S, Giannoccaro G, de Gennaro BC, Iacobellis V, Piccinni AF. Cost-benefit analysis of waste water reuse in Puglia, Southern Italy. *Water (Switzerland)* 2017; **9**(3):1–17.
29. Alcon F, Pedrero F, Martin-Ortega J, Arcas N, Alarcon JJ, De Miguel MD. The non-market value of reclaimed wastewater for use in agriculture: a contingent valuation approach. *Span J Agric Res* 2010;**8**(S2):187.
30. Verlicchi P, Al Aukidy M, Zanni G. Willingness to pay for recreational benefit evaluation in a wastewater reuse project. Analysis of a case study. *Water (Switzerland)* 2018;**10**(7):922.
31. Verlicchi P, Al Aukidy M, Galletti A, Zambello E, Zanni G, Masotti L. A project of reuse of reclaimed wastewater in the Po Valley, Italy: polishing sequence and cost benefit analysis. *J Hydrol* 2012;**432–433**:127–36.
32. Galvis A, Jaramillo MF, van der Steen P, Gijzen HJ. Financial aspects of reclaimed wastewater irrigation in three sugarcane production areas in the Upper Cauca river Basin, Colombia. *Agric Water Manag* 2018;**209**:102–10.
33. Seguí L, Alfranca O, García J. Techno-economical evaluation of water reuse for wetland restoration: a case study in a natural park in Catalonia, Northeastern Spain. *Desalination* 2009;**246**(1–3):179–89.
34. Gürlük S, Rehber E. A travel cost study to estimate recreational value for a bird refuge at Lake Manyas, Turkey. *J Environ Manage* 2008;**88**(4):1350–60.
35. Alfranca O, García J, Varela H. Economic valuation of a created wetland fed with treated wastewater located in a peri-urban park in Catalonia, Spain. *Water Sci Technol* 2011;**63**(5):891–8.
36. Ruiz-Rosa I, García-Rodríguez FJ, Mendoza-Jiménez J. Development and application of a cost management model for wastewater treatment and reuse processes. *J Clean Prod* 2016;**113**:299–310.
37. Rezaei N, Sierra-Altamiranda A, Diaz-Elsayed N, Charkhgard H, Zhang Q. *A multi-objective optimization model for decision support in water reclamation system planning*. *J Clean Prod* 2019;**240**:118227 [Internet] Available from https://doi.org/10.1016/j.jclepro.2019.118227.
38. Chhipi-Shrestha G, Rodriguez M, Sadiq R. *Selection of sustainable municipal water reuse applications by multi-stakeholders using game theory*. *Sci Total Environ* 2019;**650**:2512–26. [Internet] Available from https://doi.org/10.1016/j.scitotenv.2018.09.359.

39. Verlicchi P, Masotti L, Galletti A. Wastewater polishing index: a tool for a rapid quality assessment of reclaimed wastewater. *Environ Monit Assess* 2011;**173**(1–4):267–77.
40. Hidalgo D, Irusta R, Martinez L, Fatta D, Papadopoulos A. Development of a multi-function software decision support tool for the promotion of the safe reuse of treated urban wastewater. *Desalination* 2007;**215**(1–3):90–103.
41. Papa M, Bertanza G, Abbà A. Reuse of wastewater: a feasible option, or not? A decision support system can solve the doubt. *Desalin Water Treat* 2016;**57**(19):8670–82.

Papier | Fördert gute Waldnutzung
FSC® C083411

Kolibri 360 GmbH
Ferdinand-Jühlke-Straße 7
99095 Erfurt, Deutschland
produktsicherheit@kolibri360.de